# Deleuze and Evolutionary Theory

## *Deleuze Connections*

'It is not the elements or the sets which define the multiplicity. What defines it is the AND, as something which has its place between the elements or between the sets. AND, AND, AND – stammering.'

Gilles Deleuze and Claire Parnet, *Dialogues*



**Titles Available in the Series**

Ian Buchanan and Claire Colebrook (eds), *Deleuze and Feminist Theory*
Ian Buchanan and John Marks (eds), *Deleuze and Literature*
Mark Bonta and John Protevi (eds), *Deleuze and Geophilosophy*
Ian Buchanan and Marcel Swiboda (eds), *Deleuze and Music*
Ian Buchanan and Gregg Lambert (eds), *Deleuze and Space*
Martin Fuglsang and Bent Meier Sørensen (eds), *Deleuze and the Social*
Ian Buchanan and Adrian Parr (eds), *Deleuze and the Contemporary World*
Constantin V. Boundas (ed.), *Deleuze and Philosophy*
Ian Buchanan and Nicholas Thoburn (eds), *Deleuze and Politics*
Chrysanthi Nigianni and Merl Storr (eds), *Deleuze and Queer Theory*
Jeffrey A. Bell and Claire Colebrook (eds), *Deleuze and History*
Laura Cull (ed.), *Deleuze and Performance*
Mark Poster and David Savat (eds), *Deleuze and New Technology*
Simone Bignall and Paul Patton (eds), *Deleuze and the Postcolonial*
Stephen Zepke and Simon O'Sullivan (eds), *Deleuze and Contemporary Art*
Laura Guillaume and Joe Hughes (eds), *Deleuze and the Body*
Daniel W. Smith and Nathan Jun (eds), *Deleuze and Ethics*
Frida Beckman (ed.), *Deleuze and Sex*
David Martin-Jones and William Brown (eds), *Deleuze and Film*
Laurent de Sutter and Kyle McGee (eds), *Deleuze and Law*
Arun Saldanha and Jason Michael Adams (eds), *Deleuze and Race*
Rebecca Coleman and Jessica Ringrose (eds), *Deleuze and Research Methodologies*
Inna Semetsky and Diana Masny (eds), *Deleuze and Education*
Hélène Frichot and Stephen Loo (eds), *Deleuze and Architecture*
Betti Marenko and Jamie Brassett (eds), *Deleuze and Design*
Hélène Frichot, Catharina Gabrielsson and Jonathan Metzger (eds), *Deleuze and the City*
Colin Gardner and Patricia MacCormack (eds), *Deleuze and the Animal*
Markus P. J. Bohlmann and Anna Hickey-Moody (eds), *Deleuze and Children*
Chantelle Gray van Heerden and Aragorn Eloff (eds), *Deleuze and Anarchism*
Michael James Bennett and Tano S. Posteraro (eds), *Deleuze and Evolutionary Theory*

Visit the Deleuze Connections website at:
www.edinburghuniversitypress.com/series/delco

# Deleuze and Evolutionary Theory

Edited by Michael James Bennett
and Tano S. Posteraro

EDINBURGH
University Press

Edinburgh University Press is one of the leading university presses in the UK. We publish academic books and journals in our selected subject areas across the humanities and social sciences, combining cutting-edge scholarship with high editorial and production values to produce academic works of lasting importance. For more information visit our website: edinburghuniversitypress.com



Edinburgh University Press Ltd
The Tun – Holyrood Road, 12(2f) Jackson's Entry, Edinburgh EH8 8PJ

Typeset in 10.5/13 Adobe Sabon by
IDSUK (DataConnection) Ltd, and
printed and bound by CPI Group (UK) Ltd, Croydon, CR0 4YY

A CIP record for this book is available from the British Library

ISBN 978 1 4744 3049 4 (hardback)
ISBN 978 1 4744 3051 7 (webready PDF)
ISBN 978 1 4744 3052 4 (epub)

# Contents

# Acknowledgements

The editors would like to thank Carol Macdonald and Kirsty Woods, as well as the rest of the team at Edinburgh University Press, for their hard work and unwavering support throughout the production of this book. We would also like to express our profound gratitude for the patience and enthusiasm of our contributors. Claire Colebrook deserves thanks in particular for initiating us into the publication process and helping orient us within it.

Mike Bennett is grateful to his colleagues in the History of Science and Technology programme at the University of King's College for the first stimulus and numerous subsequent opportunities to think about these ideas. Tano Posteraro would like to thank A.Z., for everything.

# Introduction: Historical Formations and Organic Forms

*Michael James Bennett and Tano Posteraro*

Deleuze's relation to evolutionary theory is bilateral. On one side he is a legatee of the nineteenth-century tradition of evolutionism, as epitomised in the figures of Lamarck, Geoffroy Saint-Hilaire and, of course, Darwin, and as they are mediated to him by his philosophical *intercesseurs* – especially Foucault, Bergson and Nietzsche. On the other, Deleuze looks forward to new directions in evolutionary theory arising in the wake of the molecular revolution in biology. *Deleuze and Evolutionary Theory* addresses both of these faces from both a philosophical and biological point of view. It also considers some of the significance of Deleuze's and Guattari's thinking about evolution for anthropological, social and biopolitical questions.

Deleuze dreamt of a harmonious interplay between the arts, philosophy and science (1995: 125; cf. Deleuze and Guattari 1994: 208). He noted in his 1966 review of Gilbert Simondon's *L'individuation à la lumière des notions de forme et d'information* that the book 'demonstrates the extent to which a philosopher can both find his inspiration in contemporary science and at the same time connect with the major problems of classical philosophy – even as he transforms and renews those problems' (2004: 89) – a description that could just as easily apply to Deleuze's own work. Reciprocally, in defence of his and Guattari's 'utilising' of scientific references in *A Thousand Plateaus*, Deleuze asserted that 'it's not impossible for a philosopher to create concepts that can be used in science' (1995: 30). Scholars have made considerable headway in mapping out the ways in which Deleuze was 'inspired' by various aspects of biological science.[1] We focus on these two moments by way of introduction to Deleuze's career-spanning interest in evolutionary themes. Like his engagements with the history of philosophy, Deleuze's encounters with evolutionary theory begin with a relatively long 'apprenticeship' (Hardt 1993; Smith 2012), in which he comes to grips with the history of evolutionism. This is followed by a period in which Deleuze, partly in concert with Guattari, challenges what he sees as the inadequacies of historical and scientific conceptions of evolution and develops often prescient ways of thinking about it that converge with the new directions in evolutionary theory which start to emerge in the 1980s and '90s.

The restriction of Deleuze's interest in historical evolutionism to the nineteenth century is probably due to the influence of Michel Foucault. Deleuze's book-length appreciation of his friend, written shortly after the latter's death in 1984, and the appendix 'On the Death of Man and Superman' in particular, is a crucial text for understanding both Deleuze's way of thinking about the history of evolutionism and his shift to thinking about the future of biology.

The Chevalier de Lamarck is usually cited as the first modern evolutionist. His *Philosophie zoologique* (1809) argued that species change over time as a result of two forces: an inherent tendency for living bodies to increase in complexity, and the capacity of organisms to react to their changing environments. As different organs are used more extensively (or fall into disuse), they become stronger (or atrophy), and these acquired characteristics are, in turn, passed down to an organism's offspring.[2] Lamarck's evolutionary mechanism stands in contrast to Darwinian 'natural selection' (though Darwin himself accepted that the two mechanisms were not mutually exclusive) even as Lamarck and Darwin seem to be in a more fundamental agreement about the transmutation of species.[3]

In *The Order of Things*, however, Foucault argued that the 'resemblance' between Lamarck's laws and evolution in the familiar sense was only superficial. Since Lamarck conceived of the transformations of species on the basis of ontological continuity and progressive gradation, his natural history belongs to the intellectual formation of the classical age, in which evolutionism is actually impossible. The alliance between Lamarck and Darwin is an historian's ruse, which a Foucauldian genealogist ought to expose. Likewise, the narrative according to which Lamarck is a 'revolutionary' evolutionist and his long-time professional enemy the Baron Cuvier a 'reactionary' fixist is 'a fine example of simple-mindedness'. In fact, Cuvier's biology, with its four irreducibly plural *embranchements* or body-plans for the animal kingdom and its emphasis on extinction, 'introduced a radical discontinuity in the Classical scale of beings' and helped to create the intellectual conditions that made Darwinian evolution possible (Foucault 2002: 299–300; cf. 164).[4]

Deleuze's review of Foucault's book for *Le Nouvel Observateur* singles out the surprising reading of Lamarck and Cuvier (2004: 93), as does the later monograph (2006: 126). Deleuze innovates, however, by associating Foucault's historical periodisation and the difference between Lamarck and Darwin with the conceptual pair of *folding* and *unfolding*. The classical age, Deleuze says, in which Lamarck remains 'imprisoned' (2004: 93), is recognisable by the way that every force is supposed to be capable of continuous development toward infinity. Consequently, 'The *unfold* [*dépli*] appears here as a fundamental concept . . . [which] accounts for the frequency of the word "unfold" in Foucault' (2006: 126, translation modified). In contrast, by the nineteenth century, human beings have entered into relations with the 'forces of finitude', life, labour and language, and 'Everywhere it is the *Fold* [*pli*] which dominates' (128).

Deleuze's use of 'folding' as a key concept may have a biological source (Dosse 2012: 164), which happens to converge with Foucault's diction. Deleuze first discusses it in *Difference and Repetition* in the context of describing the actualisation (or differenciation) of virtual elements, as dramatised in embryogenesis (1994: 214). Deleuze is thinking of, for example, the comparative embryology of Karl Ernst von Baer. But his second reference to 'folding' appears in the context of a discussion of evolution, which he says 'we must approach . . . in terms of the pre-evolutionist polemics', such as the public debate between Cuvier and Lamarck's student Étienne Geoffroy Saint-Hilaire at the *Académie des sciences* in 1830 (215).[5] Geoffroy held that, rather than four basic *embranchements*, all animals are anatomically derived from one single structural plan, which is accessible to speculative thought if not empirical observation. Deleuze refers to Geoffroy's suggestion that the claim could be tested in terms of folding: 'is it possible to pass by folding from Vertebrate to Cephalopod?' (215; cf. 2006: 129; 1992: 165 n.25). But he notes that Geoffroy 'does not claim that the passage is carried out by folding' (1994: 215). Rather, what's essential is Geoffroy's introduction of 'developmental times which stop a given animal at a particular degree of composition' (215) – for example, an invertebrate expresses a delayed developmental stage of a vertebrate. On this basis, Deleuze, sounding Bergsonian, concludes that 'It is enough to endow time with its true meaning of creative actualization for evolution to find a principle that conditions it' (216). In other words, Deleuze presents Geoffroy as an evolutionist – or as articulating the necessary conditions for (creative) evolution – to the extent that he links the 'differenciation' of species with that of their parts. And thus the parallelism results in the famous statement that 'The entire world is an egg' (216; cf. Roffe, this volume).[6]

The significance of Geoffroy for Deleuze has been well understood by Keith Ansell-Pearson (1999: 159–60), Mark Hansen (2000), and especially Henry Somers-Hall (2012). Deleuze uses Geoffroy's transcendental approach – which makes an abstract Idea of 'Animal in itself' fundamental to zoological study – as an example of his own account of transcendental Ideas in *Difference and Repetition* (1994: 184–5). He relies particularly on Geoffroy's key concept of *homology*, according to which 'differential relations between pure anatomical elements . . . are incarnated in diverse animal configurations, with their diverse organs and functions' (185). For Geoffroy, these 'anatomical elements' are typically 'small bones' like the hyoid, homologous in human beings and cats, but functionally distinct (185; cf. Appel 1987: 149–52). In making the differential elements *abstract* – that is, parts of the 'Animal in itself' but not any particular actual animal – Geoffroy seems to anticipate Deleuze's metaphysical scheme, in which the *virtual* elements of the Idea (which are real without being actual) are actualised in time (Somers-Hall 2012: 228). But Deleuze also casts doubt on whether Geoffroy's differential elements, the 'small bones', are really 'abstract' enough. Are they not still too 'actual', too similar in kind to the animal parts into which they are differenciated? If so, then, Deleuze claims, Geoffroy's 'philosophical anatomy' is only fully realised in the science of *genetics*, in which 'genes express differential relations' and the 'whole [structure] constitutes a virtuality' – which is to say, the actualised animal and its virtual conditions don't 'resemble' one another directly (1994: 185; cf. Deleuze and Guattari 1987: 46).

Deleuze is interested in the Cuvier-Geoffroy debate from the point of view of the opposition between empiricist (actual) versus transcendental (virtual) method in science, like the German morphologists before him (Appel 1987: 158–60). From the English point of view, however, which would come to influence Charles Darwin, the significance of the debate lay in staging the conflict between *teleology*, as represented by Cuvier – the view that organic forms are determined by their functions – and the anti-teleological implication of Geoffroy's morphology. If forms precede their functions, then the two can become detached. The influence of Geoffroy on Darwin, via the anatomist Richard Owen (Ospovat 1983), led to the anti-teleological positions articulated in *The Origin of Species*. It is 'hopeless', writes Darwin, to explain the wide variety of form and function in homologous parts 'by the doctrine of final causes' (2003: 364). Darwin's debt to Geoffroy is explicit when he states that what Geoffroy had understood in terms of a transcendental unity or law of organisation, he understands as the effect of common ancestry (223–4). Darwin's evolutionary mechanism of natural selection combines such common ancestry (descent) with variations in an organism's offspring (modifications). But Darwin's rejection of teleology goes farther than Geoffroy's. Variations can be either 'injurious' or 'favourable' – which is to say, they are indifferent to any particular evolutionary outcome (144). Since there is no inherent tendency in living matter to become more complex, the complexity that does result from the struggle for existence is a genuine result of a process which selects favourable variations rather than a final cause which pre-existed it (Miquel 2007: 43). This is famously true even of the archetypical test case for natural teleology, the complex organ of the eye. Improbable though it may seem, even the eye, Darwin explains, could have been formed by the gradual accumulation of favourable but undirected variations, an argument which is immediately, and significantly, followed by a discussion about an organ in fish, the swimbladder, changing its function (2003: 211–14).

Somers-Hall emphasises the link between the decoupling of form and function inherited by Darwin and Geoffroy's 'teratology' or study of 'monsters', to which Deleuze and Guattari refer (1987: 46). While Cuvier (like the Aristotelian tradition before him) was only able to conceive of natural aberrations 'negatively' – as deviations from a plan or the *lack* of organisation – Geoffroy provides a 'positive' account of organic variation (Somers-Hall 2012: 229–30). These 'monstrous' outcomes of delayed embryogenesis are what make it possible for actualised forms to change their function (230; cf. Darwin 2003: 101–2). Somers-Hall's implication seems to be that Deleuze's metaphysics of virtual Ideas actualised in organic forms, in explicit contrast to Hegel's philosophy of nature (2012: 214–17), can be reconciled with Darwinian evolution – so long as one also takes into account the 'transversal moments' in evolutionary change such as Deleuze and Guattari discuss in *A Thousand Plateaus* (236).

We agree that Deleuze's metaphysics in the 1960s can be reconciled with, and is even sympathetic toward, features of Darwinian evolution. Deleuze's brief discussion of Darwin in *Difference and Repetition* (1994: 248) is generally positive (*pace* Eckstrand 2014), as are the allusions elsewhere (Deleuze and Guattari 1987: 47–8; Deleuze 2006: 129) – although certainly Deleuze and Guattari polemicise with 'evolution' in *A Thousand Plateaus*. It is, however, characteristic of Deleuze's fascination with genuinely Darwinian evolutionary themes, like divergence and ateleology, to develop from starting-points that are non-Darwinian or even explicitly *anti*-Darwinian. For example, Deleuze's Darwinian sympathy puts him in a somewhat awkward relation to his greatest teacher, Henri Bergson.

In *Creative Evolution* Bergson rejects the idea, associated with both Darwin and the mutationist Hugo de Vries, that organic variations are 'accidental' or due to 'chance', a point he illustrates with a memorable challenge involving the convergent evolution of eyes in mammals and marine bivalves (1998: 60–3). Bergson rejects such 'mechanism' in evolutionary biology, although he recognises the point of view as a natural one for the intellect to the extent that it is a faculty subordinated to the needs of action. Nevertheless his own evolutionism emphasises the unpredictable character of organic differenciation – perhaps even more forcefully than Darwin himself (Miquel 2007). Although explicitly a variety of 'finalism', the theory of *élan vital* presents descent and adaptation as the outcomes of a 'shared impulsion', not a progress toward a 'common aspiration' (1998: 40–1, 51–2). Unlike both mechanism and traditional finalism, Bergson's view seeks to retain the inventive, creative character of time as 'duration'. The consequences of the vital impetus are unpredetermined and unforeseeable forms (88–96).

Deleuze's 1956 essay on Bergson argues that the latter has much to contribute to a 'philosophy of difference', in part because he saw that 'Darwin helped associate the problem of difference with life, even though Darwin himself had a false conception of vital difference'. Specifically, Darwin failed to grasp that 'vital difference' is 'internal difference' (not the negative difference between two states), because natural selection, to the extent that it presupposed accidental variations of (for example) organic structure, presented differences as 'simple determinations' (Deleuze 2004: 39–40). Bergson's anti-teleological 'finalism' grasped that vital difference is, rather, 'indetermination', which means that evolution is 'unpredictable but not accidental' (Borradori 2001: 6). On this Bergsonian basis, Deleuze explains, evolution nevertheless produces the kind of divergence that Darwin predicted (2004: 40; cf. 1991: 98–9 and 1994: 212).

Another non-Darwinian starting-point for Deleuze's interest in ateleological evolution, around the time of *Difference and Repetition*, is his identification with the structuralist movement. The evolutionary and embryological dynamisms described by Geoffroy and Bergson, among others, 'incarnate differential relations immanent in Ideas' (1994: 218), which Deleuze generally describes – for example, in the eighth 'series' of *The Logic of Sense*, 'of structure' (1990: 48–51) – in terms that he himself identifies as 'structuralist'. In the contemporary essay 'How do we recognize structuralism?' Deleuze clarifies seven characteristics of structuralism, the second-to-last one of which is particularly worth flagging in connection with Deleuze's oblique appreciation of evolution. A structure 'envelops a wholly paradoxical object or element', an 'empty case', like the purloined letter in Lacan's analysis of Poe's story, or Lévi-Strauss's 'floating signifier' (2004: 184). This element lacks a proper place in any of the structure's constituent series – it is 'always displaced in relation to itself' – but in this displacement the element lends the structure movement and variation. Deleuze says it is the 'differenciator of difference' (186), language that he will later develop in *Difference and Repetition*, and in *The Logic of Sense* he identifies it with the 'event' operating as 'quasi-cause', whose 'aleatory' character he emphasises (1990: 95; cf. 56). To the extent that the movement of the 'empty case' is indeterminate, both from the point of view of correspondence with reality and conceptual entailment, this structuralist account of the event can be thought of as a theory of historical contingency (Bennett 2017: chap. 4). It is therefore unsurprising that Deleuze also links these paradoxical events with the theme of chance, which Bergson had attacked, especially in the context of discussing the 'ideal game' that 'make[s] chance into an object of affirmation' (1990: 60), yet another example of the convergence of Deleuze's metaphysics with Darwinian themes, but on distinct conceptual bases. Here the aleatory point is that of the 'dice throw' (64; cf. 1994: 198), a reference to Mallarmé, of course, but also to Nietzsche's *Zarathustra* (1969: 186; cf. Deleuze 2005: 24).

Nietzsche, like Bergson, criticised the passivity of Darwin's conception of evolution, and indeed Nietzsche is another anti-Darwinian source for Deleuze's Darwinian sympathy. In his reading of Nietzsche's 'theory of forces', Deleuze notes that Nietzsche diagnosed the 'taste' of modernity – and of modern science in particular – for reactive, or dominated, forces rather than active, dominating ones (2005: 38, 42). Deleuze derives the distinction from the passage in *The Genealogy of Morals* on Will to Power (II.12), which criticises how the 'democratic idiosyncrasy which opposes everything that dominates and wants to dominate' has robbed the theory of life of its 'fundamental concept', activity, and 'place[d] instead "adaptation" in the foreground, that is to say, an activity of the second rank, a mere reactivity' (Nietzsche 1989: 78–9). Nietzsche's own view is that 'all events in the organic world are a subduing, a *becoming master*' – that is, involve an active (dominating) force – and he memorably identifies this 'subduing' with giving a 'fresh interpretation . . . through which any previous "meaning" and "purpose" [of, for example, an organ] are necessarily obscured or even obliterated' (77). The detachment of organic form and function, which permits putting organs to new purposes, is, however, a genuinely Darwinian idea, inherited ultimately from Geoffroy, and the evolutionary mechanism that 'reinterprets' organs, called 'exaptation', is now usually thought to be compatible with a generous Darwinism (Gould and Vrba 1982; cf. Richardson 2004: 44–5).

Likewise, Deleuze sounds Darwinian (Caygill 1997) when he stipulates that the 'becoming active' of forces is the product of a 'selection' performed, famously, by the thought of the Eternal Return, which 'eliminate[s] certain reactive states' and their bearers (like the 'small reactive man'), leaving only the Overman (2005: 63–4; cf. 1994: 8). Deleuze's recognition of Nietzsche's explicit preference for Lamarck (2005: 39) is therefore ironic; Deleuze is more Darwinian here than Nietzsche, who explicitly denies that a 'selective principle' like the one involved in natural selection would favour the strong and active type (1997: 59) and derides the 'scholarly oxen' who see a link between the theme of the Overman and a version of Darwinism (1989: 261).

Deleuze returns to the 'theory of forces' in the appendix to *Foucault* where his reception of historical evolutionism comes full circle. After discussing the intellectual forms produced by the forces at work in the nineteenth century, Deleuze attempts to use Foucault's historical periodisation to make predictions about 'formations of the future' (2006: 129–31) and cautiously hints again at the theme of the Overman. Future intellectual formations will be linked to contemporary developments in machine technology and literature, he says, but, once again, biology is Deleuze's favoured model: 'Biology had to take a leap into molecular biology, or dispersed life regroup in the genetic code' (131). In technology, literature and biology, the Overman (*surhomme*) of the future will affirm what Deleuze calls the '*Superfold* [*surpli*, or "overfold"]', which is neither infinite nor finite but an 'unlimited finity [*fini-illimité*]', in which 'a finite number of components yields a practically unlimited diversity of combinations', the paradigm case being the folding of the genetic code (131). Deleuze predicts that the distinction between forces intrinsic to the organism ('man') and extrinsic to it will break down – which happens, for instance, when the 'genetic components which supersede the organism' are subject to lateral transfers between systems or 'captures' of code (132).

This appendix thus forms a hinge. Deleuze turns from considering the history of evolutionary science, as it is mediated to him philosophically, to address explicitly the future of biology and the new directions for evolutionary theory beyond the inherited mechanisms of descent with modifications.

It comes as no surprise, then, that the character of Deleuze's engagement with biological themes begins to shift, especially in his collaborative work with Guattari. Deleuze's remarks on the life sciences come to be marked by a newly polemical edge. It is, however, a significant part of the argument of this volume that *Capitalism and Schizophrenia*'s radically interdisciplinary and experimental creation of concepts evinces an exciting and excited – even if fractional and oblique – engagement with the cutting-edge biology of the time. No doubt, *A Thousand Plateaus* is studded with dismissive criticisms of linear 'evolutionism' and the arboreal model of evolution by descent with modification; Deleuze and Guattari even define a set of their central concepts *against* the evolutionary-theoretic mainstays of heredity, filiation and genealogy, going so far as to promote 'involution' as an alternative term so as to definitively distinguish evolutionary understandings of life from their own (1987: 238–41). Yet we maintain that these comments should be taken not as dismissals of evolutionary theory *tout court*, but as path-breaking anticipations of the new directions that theory and its concomitants was just then on the verge of taking.

*A Thousand Plateaus* – published in 1980 – was, on this front, nothing short of prescient. It presaged, in a number of ways, what was to be a veritable sea change in the study of life. Over the next several decades evolutionary biology was shaken with a series of shockwaves that would culminate in what theorists today are starting to call an 'Extended' and even 'Postmodern' Evolutionary Synthesis in order to mark it off from the Modern Synthesis still dominant in the 1980s (Laland et al. 2015; Brucker and Bordenstein 2014). The hologenome theory of evolution – which posits that the organism and its microbiome form a single, multifaceted unit of selection – signals perhaps better than any other theoretical development the advent of a new evolutionary synthesis. This is a synthesis founded in the increasingly important role afforded to symbiosis in explanations of the evolution of cells and species, from the endosymbiotic relationships responsible for the genesis of mitochondria from out of ancient bacterial alliances to the phylosymbiotic relationships between host species and their associated microbial communities constitutive of speciation as such (Brooks et al. 2017). One important consequence of these developments is a newfound appreciation for horizontal gene transfer and aparallel evolution in microorganisms, as well as cytoplasmic, environmental, behavioural and symbolic forms of transmission in almost everything else (Rosenberg and Zilber-Rosenberg 2016).

These 'new directions' work to disrupt and disperse two distinctions long taken to be self-evident and indispensable to evolutionary theory: between the organism and its genes, and between the organism and its environment. The critique and complication of each finds its positive complement in a new set of developments as well. Criticisms of genetic reductionism and the pan-adaptationist programme correspond to the elaboration of decentralised accounts of causality and transmission in Developmental Systems Theory and 'Evo/Devo' (Oyama 2000); and criticisms of autonomous biological individuals, distinct genetic lineages and strictly vertical models of inheritance have developed alongside a newfound appreciation for the ubiquity of symbiosis, microbial alliances and horizontal gene transfer both as features of constituted organisms and as a source for variation and evolutionary discontinuities in the history of life as well (Gilbert et al. 2012).

Deleuze and Guattari appear to have been ahead of the curve on a number of these counts. To begin, *Difference and Repetition*'s embryologically articulated theory of individuation is constructed in no small part on the basis of a rejection of genetic determinism and reductionism. Deleuze argues there that while the ontogenetic process originates in a 'pre-individual' domain structured by multiply actualisable differential relations among nucleic and genetic elements, 'their actualization is determined only by the cytoplasm, with its gradients and its fields of individuation' (1994: 251). The expression of genetic information, in other words, is developmentally enabled and constrained by a differential network of both genetic and epigenetic factors. And as the differential relations among these factors are to be modelled as virtual – at least according to the metaphysics of *Difference and Repetition* – they are therefore open to an unpredetermined set of actualisations, one that fluctuates correlative with the contingencies of development. As the developmental process is sensitive to environmental factors as well, the gene is best understood as an abstraction from a set of relations structuring a field whose outcome is to be determined over the course of its own development, with all the plasticity and unforeseeability that implies. There is, as several commentators have already indicated, a striking overlap between this model of ontogenesis and the accounts furnished by more recent work in Developmental Systems Theory (Protevi 2013: 197–211; Ansell-Pearson 1999: 144, 157, 165–6).

When it comes to the distinction between the organism and its environment, Deleuze and Guattari also seem to have anticipated many tenets of what would come to be called Niche Construction Theory (Odling-Smee et al. 2003). Their concept of the 'haecceity' blurs the demarcating lines between 'climate, wind, season, hour' and the organisms that populate them in order to argue that each element plays a constructive role in the ongoing and fluctuating composition of each other (Deleuze and Guattari 1987: 263). Organisms, on this account, do not evolve and interact against an environmental backdrop independent from them, but rather come to be as nodal points in the 'haecceity', or individuation, of a larger assemblage spanning their bodies and activities and the bodies and activities around them, both organic and not (262). Just as the gene was positioned in *Difference and Repetition* as an abstraction from a network of reciprocally related elements, so is the individual organism similarly reconceptualised in *A Thousand Plateaus* as an abstraction from the assemblages and 'blocks of becoming' that 'snap it up' and in which it develops (238).

Symbiosis is perhaps above all the most important concept in Deleuze and Guattari's simultaneous rebuke of evolutionary orthodoxy and insight into the new directions it was soon to take. It functions negatively as a critique of the descent-based filiative model of evolution: 'becoming', as Deleuze and Guattari so memorably put the point, 'is not an evolution, at least not an evolution by descent and filiation' (238). They add that 'if evolution includes any veritable becomings, it is the domain of *symbioses* that bring into play beings of totally different scales and kingdoms, with no possible filiation' (238). Symbioses span kingdoms, dislocate and propagate genetic information across independent evolutionary lineages, afford new capacities to otherwise less able individual organisms, and take up each of their terms in a co-evolutionary process that leaves them not only transformed together, but indissociable from each other as well.

Lynn Margulis, the theorist responsible for popularising the concept, was a proponent of the counter-Darwinian thesis that evolution operates *primarily* through horizontal communal relations (Margulis and Sagan 1997: 189–90). 'Descent with modification' is derivative. After confirming that the cellular mitochondria of today were synthesised about two billion years ago out of formerly free-living bacteria – what she calls 'symbiogenesis' – Margulis went on to argue that *all* seemingly individual organisms are products of similar events, that is, that all organisms 'originated by symbiont integration that led to permanent associations' (Margulis and Sagan 2002: 6, 142).

Today this insight has ramified in a complex array of directions, a crucial one of which is the discovery that macro-scale organisms develop, function, reproduce and evolve through the environmental acquisition of a large number of microbial symbionts. The much-celebrated 'microbiome' designates the set of microbes associated with a host organism; the still-controversial 'holobiont' designates the biological unit composed of the organism together with its microbial symbiotic community; and the 'hologenome' is made up of the collective genomes of all of them

(Gilbert et al. 2012). These developments have rebounded on just about every tenet of evolutionary theory in the form of questions, problems and puzzles, not least of which is how we are to conceptualise a biological individual if it is indissociable from a dazzlingly large set of genetically distinct microbes and evolves with them, despite not transmitting them to its offspring hereditarily.

Here, too, Deleuze and Guattari appear prescient. Symbioses are for them a kind of assemblage, one whose components, for the most part, are made up of organisms interacting closely in a way that significantly modifies the lifecycles of some of them (cf. Allen, this volume). But the concept of the assemblage is wider than that of the symbiotic unit. Assemblages are made of enduring coordinations of heterogeneous components of any type, concretised or actualised on the basis of an 'abstract machine', which is like the virtual structure presiding over the development of the egg in *Difference and Repetition*. In this case, it is a kind of diagram consisting of a set of differential relations between elements, the various functions of which they are capable and their potential material incarnations (Deleuze and Guattari 1987: 141). It is by remaining in contact with its abstract machine that the concrete assemblage can be modified, swap out its parts, and manifest new functions. The abstract machine is a reservoir of potentialities from which the concrete assemblage draws; it is also what structures or formats it across changes in its components. The organism might be understood as an assemblage in just this sense: it consists of a coordination among various parts sourced from elsewhere, acquired both vertically, by heredity, and horizontally, through its integration in an environmental haecceity; and it is structured on the basis of an abstract diagram that outlines possible parts and the functions that would enlist them. Modelled this way, the organism as assemblage anticipates the holobiont, and the abstract machine on whose basis the assemblage is effectuated and concretised anticipates the turn from the genome to the hologenome as well.[7]

At a phylogenetic scale, descent-based Darwinism is also in the process of being updated according to what Deleuze and Guattari envisioned as a 'rhizomatic' model of 'transversal' connections, the effect of which is the gradual replacement of the genealogical tree and its linear, branching lines of speciation with a network or 'rhizome' composed of microbial alliances, 'jumping genes' and trans-lineal symbioses (Suh et al. 2015; cf. Raoult 2010). The change in model reflects what is increasingly considered to be the essentially hologenomic and trans-specific bases of speciation events. The orthodox model is 'arborescent': it conceives differentiation through shared lines that branch outward as they develop, forming the shape of a tree – as in Darwin's famous illustration in *On the Origin of Species*. It isolates and restricts the interactions and events capable of producing the emergence of novelty to separate evolutionary trajectories. But nature can now be seen to produce otherwise, through cross-connections that span different phyla, aparallel co-evolutionary feedback loops, reciprocal 'deterritorialisations' or dislocations of parts from functions and their re-specification in other terms in tandem, and heterogeneous assemblages or quasi-individual coordinations of genetically unrelated entities (Deleuze and Guattari 1987: 10). 'Transversal communications between different lines', as Deleuze and Guattari foresaw, 'scramble the genealogical trees' (11). Viruses spread from the cellular material of one species into another, infecting the second with traces of information from the first (10). Orchids deterritorialise by forming an image of wasps, and wasps reterritorialise onto those images, but at the expense of deterritorialising themselves and becoming components in the orchids' reproductive apparatuses (10). The wasp and orchid, reciprocally deterritorialised, form a rhizome and develop in tandem as such. Deleuze and Guattari call this kind of creative transformation 'involutionary', again not to mark it off as unintelligible in an evolutionary register, but to emphasise the difference between orthodox evolutionism's foundation in filiation and descent and their machinic account of heterogeneous assemblages, symbiotic alliances, transversal communications and contagion (238).

These are today indispensable insights. Lynn Margulis's long-neglected work on the endosymbiotic genesis of eukaryotic cells from out of alliances forged between primitive prokaryotic cells is inspiring a new wave of interest in symbiosis and horizontal gene transfer as effective evolutionary events. Mary Jane West-Eberhard's theory of developmental plasticity has made it increasingly difficult to attribute direct causal influence to almost any one set of factors in the development of the living thing. And Scott Gilbert's theory of the holobiont more or less directly confirms the move away from the bounded organism toward the open-ended assemblage whose variable component parts span organismal, genetic, viral, parasitic and symbiotic registers. Finally, we might consider too the way that the endeavour to 'make sense of life' can today no longer be dissociated from the various instruments, technologies, models, metaphors and programs through which biological practice is now conducted (Keller 2002: 2–3). An increasingly unwieldy variety of entities are brought into circulation around even the most ostensibly basic biological postulate: the gene, for instance, is always instantiated in technological assemblages and makes little sense abstracted from them. Recent theorists have drawn two conclusions: first, that the models, machines and metaphors that have contributed so forcefully to the progress made in recent biology have also worked to undermine its unity, autonomy and completeness by opening the study of life onto informatics, modelling and manipulation; and second, that these new assemblages 'leave the project of "making sense of life" with an essentially – and perhaps necessarily – mosaic structure' (2–3). We can see too, finally, in Deleuze and Guattari's rhizomes and assemblages, striated as they are across a series of natural and ontological registers, the anticipation of life's irreducibly 'mosaic' structure, a structure that is steadily outstripping orthodox evolutionism's theoretical resources.

The chapters collected here span three categories in the relationship between Deleuze and evolutionary thought. Barry Allen, Jon Roffe and Erin Hortle with Hannah Stark each interrogate the *philosophical* dimensions of Deleuze's debt to, critical engagement with, and appropriations of the biology and evolutionary theory of his time, while Michael Bennett and Paul-Antoine Miquel uncover and reconstruct the complex *biological* affinities, anticipations and divergences that link Deleuze's thinking with recent research in the history and philosophy of the life sciences. Johan Normark, Daniel W. Smith, and Claire Colebrook conclude the volume by considering some of the *social*, *political* and *anthropological* facets of the Deleuze-evolution connection, from the implication of life within the ideas and forms of its politicisation, to the framing of archaeological and anthropological issues in Deleuzo-Guattarian terms, as well as the broader question of the logic and operation of evolutionary processes on the level of the social in general.

Barry Allen's 'Unnatural Nuptials' opens the volume by tracking the evolution of the concept of life from *Difference and Repetition* through *A Thousand Plateaus*. Deleuze's engagement with biology begins in the first text with Darwin, whose thought of individual difference reversed the priority accorded since Aristotle to the species and re-centred it on particular organisms and their varying characteristics. *A Thousand Plateaus* infamously decentred the primacy accorded to the organism by deriving it from the genetic line continuous through it and reconceiving it as an instance of the populations that code it. But most interesting, for Allen, is the way coded individuals are relocated within the assemblages and symbiotic alliances that comprise the real loci of the evolutionary movement. These alliances make for the 'unnatural nuptials' from which Allen's chapter takes its name and which operate, for Deleuze and Guattari, as 'the true Nature spanning the kingdoms of nature' (1987: 241).

Jon Roffe's 'The Egg: Deleuze Between Darwin and Ruyer' refocuses Deleuze's engagement with biology by problematising *Difference and Repetition*'s characterisation of Darwin. Roffe argues that what might appear at first as a creative misappropriation of Darwin has to be understood through the supplementation of the embryological line of thought that runs through *Difference and Repetition* and inspires its account of ontogenesis. Roffe prosecutes this argument by reconstructing what Deleuze takes from the sequence of embryologists that appears after his invocation of Darwin and culminates with Raymond Ruyer. Roffe holds that it is not Darwin but Ruyer who serves as the key reference point for Deleuze's elaboration of a biology-based account of individuation.

In 'Framing Sexual Selection: Elizabeth Grosz's work on Deleuze, Darwin and Feminism', Erin Hortle and Hannah Stark take up the relationship between Deleuze and evolution from a newly critical angle, by thinking through the position of sexual difference in Deleuze's appropriation of Darwin. They thematise this issue by routing it through a study of the work of Elizabeth Grosz, whose own thought has unfolded through a complex confrontation between Luce Irigaray's ethics of sexual difference and Deleuze's metaphysics of difference as such. The tension between the latter two is clear enough: difference is to be thought, for Deleuze, prior to the genesis of any identity, category, or distinction, whereas sexual difference – the difference between masculinity and femininity – is, for Irigaray, ontologically irreducible. Hortle and Stark hold that it is Grosz's use of Darwin that allows her to mediate between Irigaray and Deleuze. Grosz agrees with Irigaray that sexual difference drives (and accelerates) the production of differences while also maintaining with Deleuze that sexual difference was itself produced through more general processes of differentiation. Grosz secures the latter position by way of her reading of Darwin, which privileges two facets of his account of evolution by natural selection: first, the idea that nature selects for sexual difference; and second, the idea that sexual difference has as its evolutionary correlate sexual selection, which functions as an engine of creativity operative within and alongside natural selection. Grosz is best understood, then, for Hortle and Stark, as grounding Irigaray's ethics of sexual difference in Deleuze's differential ontology, rethought through Darwin's theories of natural and sexual selection. The upshot is a philosophy of difference, and a concomitant conception of nature, that centres on sexual difference and its ethico-political implications.

Michael J. Bennett's chapter sets out to assess the comparisons often made by Deleuze's readers between his work and recent trends in the philosophy of biology – in particular the 'Developmental Systems Theory' endorsed by Susan Oyama and others. On Bennett's view, Deleuze and Guattari are actually out of step with contemporary advocates of 'DST' to the extent that they tend to privilege the developmental role of genes. The comparison is, however, illuminating in terms of the disciplinary division of labour between science and philosophy, and it illuminates Deleuze and Guattari's perplexing discussion of the two in *What is Philosophy?* While Oyama shows how avoiding misleading metaphors like 'genetic blueprint' can make for better scientific descriptions, her perspective has been criticised as hindering scientific practice, which often deploys heuristics or metaphorical models in research that are, strictly speaking, inadequate or even false. As a result, DST has been called a contribution not only to scientific research but to the 'philosophy of nature', a related but distinct endeavour. Bennett argues that this distinction also clarifies Deleuze and Guattari's distinction between science and philosophy. While the former deploys useful, but potentially false, heuristics, the latter attempts to do without them and to approximate the 'infinite speed' of becoming. According to Bennett, this interpretation makes good sense of Deleuze's wariness about *metaphor*, a suspicion he shares with Oyama, and implies that Deleuze and Guattari's accounts of embryogenesis and symbiosis may be described as a kind of philosophical holism – so long as it is appropriately distinguished from the implicitly Hegelian varieties that privilege internal relations.

Paul-Antoine Miquel's 'Deterritorialisation and Creative Involution' also addresses the question of Deleuze and Guattari's scientific viability, making a case for it by reframing their theory of non-organic life by way of reference to Lovelock and Margulis. Miquel begins accordingly with Deleuze and Guattari's concept of the Earth, which he understands as an ontological principle of foundation without ground whose operative logic is rhizomatic. The Earth, understood rhizomatically, is always less than itself, non-totalisable, structurally incomplete. Lines of relative deterritorialisation populate each of its strata. Miquel argues that the organic stratum, or biosphere, is not so much independent of the physicochemical stratum, or geosphere, as it is constituted by the lines of flight or relative deterritorialisations of the latter. The biosphere is conditioned by the geosphere and regulates the chemical conditions of the geosphere in turn – through feedback, like an enormous thermostat. It is this transversal line, which departs from the geosphere in order to run through the stratum of biological organisation where it regulates and re-founds (without grounding) the geosphere itself, that Miquel calls non-organic life. It is 'enveloped in the geosphere', as he puts it, 'and developed in the biosphere'. Life – recalling Bergson's formulation – exceeds its delimitation within individual organisms. Miquel argues that non-organic life can be made to accommodate the recent challenges put to the biological individual by symbiosis as well as those put to orthodox Darwinism by horizontal gene transfer.

Johan Normark's chapter, 'Hydrosocial Becomings: Evolutionary Perspectives on Water Assemblages and Maya Kingship', shifts the collection's focus from biological to anthropological questions and to the possibility of social or cultural evolution. Normark deploys a Deleuzo-Guattarian perspective to critique the cultural evolutionism prevalent in his own field of archaeology, particularly the Darwinian- and neo-Darwinian-inspired perspectives that treat cultures as super-organisms and cultural items rather than genes as units of inheritance. His analysis of the significances of water and kingship in Classical Mayan cultures builds on the work of Manuel DeLanda and Levi Bryant, who extend Deleuze's discussion of the actualisation of a virtual structure beyond the embryological register to develop theories of multi-scalar assemblages and exo- and endo-relations among machines that cut across the distinctions between nature, culture, life and technology. This approach generates the conclusion that, in some respects, the 'machine' of kingship among the Classical Maya was less robust than, and revolved around, the machines for water management and the domestication of maize, a result that argues for the value of the Deleuzo-Guattarian approach to cultural evolution.

In contrast, Daniel W. Smith's chapter argues that in the two volumes of *Capitalism and Schizophrenia* Deleuze and Guattari expose the shortcomings of evolutionary thinking in the political and anthropological register as starkly as in the biological. They are 'Against Social Evolution', at least in the sense that their analysis of ancient despotic states like Babylon and Egypt denies the possibility that such complex states could have emerged or evolved out of 'primitive' hunter-gatherer cultures. Clarifying many of Deleuze and Guattari's archaeological and anthropological influences – like Marshall Sahlins and Jane Jacobs via Pierre Clastres and Fernand Braudel – Smith explains the 'antinomy' involved in the traditional accounts of the origins of the state – for instance, the Marxist account – and Deleuze and Guattari's alternative concept of the 'apparatus of capture'. Deleuze and Guattari's anti-evolutionary socio-political thought entails some surprising implications – for example, that the apparatus does not simply appropriate something that existed before and independently of the State, but rather *creates* what it captures, as well as the counterintuitive belief that there has only ever been one state. Smith's unpacking of these claims draws on Deleuze's metaphysics of time, originally derived from the philosophy of biology (Bergson and Ruyer), according to which time is not a chronological movement from one actualised state to another (a 'counting of motion', as Aristotle thought) but an actualisation of virtual Ideas. According to Smith, Deleuze and Guattari's *Urstaat* is one such Idea, the diverse actualisations of which in historical formations account for the multifarious becomings in socio-political history. In anthropology, as in biology, 'becoming is not an evolution' (Deleuze and Guattari 1987: 239).

Finally, Claire Colebrook rounds out the collection by stepping back and reflecting on the question of the relationship between philosophy and life, which underpins both much of post-phenomenological philosophy and contemporary biopolitics. Does philosophy ultimately transcend the topic of life and provide a basis from which to criticise or legislate over its deployment, or does a scientific understanding of life inform philosophical notions of mind, matter and even ethics? Colebrook aims to trouble this seemingly exhaustive opposition by finding a third possibility in the work of Deleuze and Guattari. The distinct relations between philosophy and life presupposed by readers generate parallel tendencies in their readings of Deleuze and Guattari. Some interpret them as advocating a vital, embodied materialism, in which supposedly organic wholes are revealed as ongoing dynamic formations constantly in contact with exterior milieus, and others as emphasising the power of thought beyond life, a radical break from ordinary vital processes like self-maintenance. Colebrook, in contrast, theorises what she calls 'radical' or 'general epigenesis', epitomised in Deleuze and Guattari's descriptions of deterritorialisation and stratification. This means the complete collapse of the distinction between interior and exterior, which Colebrook shows to be retained in various vestigial ways by contemporaries of Deleuze – like Foucault and Derrida – and more recent thinkers, like Meillassoux, Badiou and Stiegler. Colebrook reveals how, counterintuitively, Deleuze develops his radical epigenesis in his book on Leibniz in terms of the doctrine of *preformation*, with which it is usually contrasted. And most crucially she suggests that rethinking the link between philosophy and life – thinking of it in terms of implication and explication rather than 'relation', which implies too much of an internal/external conditioning – creates new possibilities for biopolitics and for an 'ethics of life'.

## Notes

1. For example, Keith Ansell-Pearson (1999) has insisted on the centrality of an encounter with the neo-Darwinism of August Weismann to Deleuze and Guattari's 'biophilosophy'. Howard Caygill (1997) has criticised Deleuze's emphasis on the

theme of population 'distribution' over the immanent rigour of Darwinian 'selection', while Elizabeth Grosz has placed Darwin, Bergson and Deleuze together in an 'indirect filiation' on the basis of their shared emphasis on the indeterminacy and contingent self-organisation of life (2011: 34). Todd May (2005) and John Marks (2006) have illuminated the debt Deleuze owes to François Jacob and Jacques Monod.

2. On Lamarck's views and the relevance of the inheritance of acquired characteristics for contemporary biology, see Jablonka and Lamb 1995.
3. For Darwin's pluralism about evolutionary mechanisms, see Darwin 2003: 180, and on the contrast between the two belying a deeper agreement, Perrier 1884: 78–9.
4. For a critique of Foucault's view, see Bowler 2009: 93–5. Jablonka and Lamb (1995) also emphasise the extent to which Lamarck predicts evolutionary divergence and challenges the progressive continuum of life.
5. The debate between Geoffroy and Cuvier has been subject to multiple, sometimes mythologising, reinterpretations, but it is particularly characteristic of the late nineteenth century in France to treat it as an anticipation of (and proxy for) the controversy over descent with modifications (Appel 1987: 233–5). The zoologist Edmond Perrier (1884), on whom Deleuze relies (1994: 216, 219; Deleuze and Guattari 1987: 522 n.8), was especially committed to this construal.
6. Strangely, in the *Foucault* book, Deleuze says that 'What Cuvier, Geoffroy, and Baer also have in common [in addition to "folding"] is that they resist evolutionism' (2006: 129). In fact, Geoffroy's variant of the Lamarckian evolutionary mechanism emphasised the power of the environment to act on the developing embryo, thus producing 'monstrosities'. As a consequence, Geoffroy's account has been called a 'teratological' theory of evolution (Appel 1987: 130–5) and an ancestor of 'saltative' theories, according to which new species emerge by abrupt jumps (Bowler 2009: 261).
7. It bears noting, however, that Deleuze and Guattari are themselves hostile to the organism throughout *A Thousand Plateaus*. Nonetheless, there remain opportunities for its reconceptualisation – in a form more amenable to Deleuze and Guattari's own commitments and antipathies – according to recent developments in the scientific determination of biological individuals.

## References


Ansell-Pearson, K. (1999), *Germinal Life: The Difference and Repetition of Deleuze*, London: Routledge.

Appel, T. A. (1987), *The Cuvier-Geoffroy Debate: French Biology in the Decades Before Darwin*, New York: Oxford University Press.

Bennett, M. J. (2017), *Deleuze and Ancient Greek Physics*, London: Bloomsbury.

Bergson, H. (1998), *Creative Evolution*, trans. A. Mitchell, Mineola, NY: Dover.

Borradori, G. (2001), 'The Temporalization of Difference: Reflections on Deleuze's Interpretation of Bergson', *Continental Philosophy Review* 34, pp. 1–20.

Bowler, P. J. (2009), *Evolution: The History of an Idea, 25th Anniversary Edition*, Los Angeles: University of California Press.

Brooks, A., K. Kohl, R. Bruckner, E. van Opstal and S. Bordenstein (2017), 'Phylosymbiosis: Relationships and Functional Effects of Microbial Communities across Host Evolutionary History', *Public Library of Science: Biology* 15: 1.

Brucker, R. and S. Bordenstein (2013), 'The Hologenomic Basis of Speciation: Gut Bacteria cause Hybrid Lethality in the genus Nasonia', *Science* 341: 6146, pp. 667–9.

Caygill, H. (1997), 'The Topology of Selection: The Limits of Deleuze's Biophilosophy', *Deleuze and Philosophy: The Difference-Engineer*, ed. K. Ansell-Pearson, London: Routledge, pp. 149–62.

Darwin, C. (2003), *On the Origin of Species*, ed. J. Carroll, Peterborough, ON: Broadview Press.

Deleuze, G. (1990), *The Logic of Sense*, trans. M. Lester with C. Stivale, New York: Columbia University Press.

Deleuze, G. (1991), *Bergsonism*, trans. H. Tomlinson and B. Habberjam, New York: Zone Books.

Deleuze, G. (1992), *The Fold: Leibniz and the Baroque*, trans. T. Conley, Minneapolis: University of Minnesota Press.

Deleuze, G. (1994), *Difference and Repetition*, trans. P. Patton, New York: Columbia University Press.

Deleuze, G. (1995), *Negotiations: 1972–1990*, trans. M. Joughin, New York: Columbia University Press.

Deleuze, G. (2004), *Desert Islands and Other Texts (1953–1974)*, ed. D. Lapoujade, trans. M. Taormina, Los Angeles: Semiotext(e) Foreign Agents Series.

Deleuze, G. (2005), *Nietzsche and Philosophy*, trans. H. Tomlinson, New York: Continuum.

Deleuze, G. (2006), *Foucault*, trans. S. Hand, Minneapolis: University of Minnesota Press.

Deleuze, G. and F. Guattari (1987), *A Thousand Plateaus: Capitalism and Schizophrenia*, trans. B. Massumi, Minneapolis: University of Minnesota Press.

Deleuze, G. and F. Guattari (1994), *What is Philosophy?*, trans. H. Tomlinson and G. Burchell, New York: Columbia University Press.

Dosse, François (2012), *Gilles Deleuze and Félix Guattari: Intersecting Lives*, trans. D. Glassman, New York: Columbia University Press.

Eckstrand, N. (2014), 'Deleuze, Darwin, and the Categorisation of Life', *Deleuze Studies* 8: 4, pp. 415–4.

Foucault, M. (2002), *The Order of Things: An Archaeology of the Human Sciences*, London: Routledge.

Keller, E. Fox (2002), *Making Sense of Life: Explaining Biological Development with Models, Metaphors, and Machines*, Cambridge, MA: Harvard University Press.

Gilbert, S., J. Sapp and A. Tauber (2012), 'A Symbiotic View of Life: We Have Never Been Individuals', *The Quarterly Review of Biology* 87: 4, pp. 325–41.

Gould, S. J. and E. S. Vrba (1982), 'Exaptation – A Missing Term in the Science of Form', *Paleobiology* 8: 1, pp. 4–15.
Grosz, E. (2011), *Becoming Undone: Darwinian Reflections on Life, Politics, and Art*, Durham, NC: Duke University Press.
Hansen, M. (2000), 'Becoming as Creative Involution? Contextualizing Deleuze and Guattari's Biophilosophy', *Postmodern Culture* 11: 1, <http://pmc.iath.virginia.edu/text-only/issue.900/11.1hansen.txt> (last accessed 30 July 2018).
Hardt, M. (1993), *Gilles Deleuze: An Apprenticeship in Philosophy*, Minneapolis: University of Minnesota Press.
Jablonka, E. and M. J. Lamb (1995), *Epigenetic Inheritance and Evolution: The Lamarckian Dimension*, New York: Oxford University Press.
Laland, K., T. Uller, M. Feldman, K. Sterelny, G. Müller, A. Moczek, E. Jablonka and J. Odling-Smee (2015), 'The Extended Evolutionary Synthesis: Its Structure, Assumptions and Predictions', *Proceedings of the Royal Society B: Biological Sciences* 282: 1813.
Margulis, L. and D. Sagan (1997), *Microcosmos: Four Billion Years of Microbial Evolution*, Berkeley: University of California Press.
Margulis, L. and D. Sagan (2002), *Acquiring Genomes*, New York: Basic Books.
Marks, J. (2006), 'Molecular Biology in the Work of Deleuze and Guattari', *Paragraph* 29: 2, pp. 81–97.
May, T. (2005), *Gilles Deleuze: An Introduction*, New York: Cambridge University Press.
Miquel, P.-A. (2007), 'Bergson and Darwin: From an Immanentist to an Emergentist Approach to Evolution', *SubStance* 36: 3, pp. 42–56.
Nietzsche, F. (1969), *Thus Spoke Zarathustra*, trans. R. J. Hollingdale, New York: Penguin Books.
Nietzsche, F. (1989), *On the Genealogy of Morals and Ecce Homo*, ed. W. Kaufmann, New York: Vintage.
Nietzsche, F. (1997), *Twilight of the Idols*, trans. R. Polt, Indianapolis: Hackett.
Odling-Smee, J., K. Laland and M. Feldman (2003), *Niche Construction: The Neglected Process in Evolution*, Princeton: Princeton University Press.
Ospovat, D. (1983), *The Development of Darwin's Theory: Natural History, Natural Theology, and Natural Selection 1838–1859*, New York: Cambridge University Press.
Oyama, S. (2000), *Evolution's Eye: A Systems View of the Biology-Culture Divide*, Durham, NC: Duke University Press.
Perrier, E. (1884), *Philosophie zoologique avant Darwin*, Paris: Alcan.
Protevi, J. (2013), *Life, War, Earth: Deleuze and the Sciences*, Minneapolis: University of Minnesota Press.
Raoult, D. (2010), 'The Post-Darwinist Rhizome of Life', *Lancet* 375: 9709, pp. 104–5.
Richardson, J. (2004), *Nietzsche's New Darwinism*, New York: Oxford University Press.
Rosenberg, E. and I. Zilber-Rosenberg (2016), 'Microbes Drive Evolution of Animals and Plants: the Hologenome Concept', *American Society for Microbiology* 7: 2.

Smith, D. W. (2012), 'Deleuze and the History of Philosophy', *The Cambridge Companion to Deleuze*, ed. D. W. Smith and H. Somers-Hall, New York: Cambridge University Press, pp. 13–32.

Somers-Hall, H. (2012), *Hegel, Deleuze, and the Critique of Representation: Dialectics of Negation and Difference*, Albany: State University of New York Press.

Suh, A., L. Smeds and H. Ellegren (2015), 'The Dynamics of Incomplete Lineage Sorting across the Ancient Adaptive Radiation of Neoavian Birds', *PLOS Biology* 13: 8.

# Chapter 1

# Unnatural Nuptials

*Barry Allen*

In *Difference and Repetition*, Charles Darwin was the philosopher of individuals and the priority of individual differences. His theory of evolution inaugurated 'the thought of individual difference' (Deleuze 1994: 248). For Darwin, the individual and its differences come first. Species-specific characteristics do not exist until natural selection does its work on individual differences over geological time. It is not the individual that is derivative in relation to the genus of the species, which was the conclusion of Aristotelian science. Rather, it is the species that 'is an illusion – inevitable and well founded, it is true – in relation to the play of the individual and individuation. . . . It is the individual which is above the species and precedes the species in principle' (250).

Twelve years later, in *A Thousand Plateaus*, a neo-Darwinian theme intrudes. Individuals are described as effects of populations. Species-identity ('form') depends on codes, and codes apply only to populations, not individuals. 'Organic form depends on an autonomous code', a neo-Darwinian mantra Deleuze and Guattari repeat four times in less than ten pages (1987: 51, 52, 54, 59). Individuals are coded because they emerge amid populations that are coded, and changes in code primarily affect populations by affecting the ability of the code to propagate, which requires the agency of individual organisms. No code is impervious to decoding, organically expressed as mutation, genetic drift and the re-purposing of redundancy. Fragments of code transfer across forms symbiotically. All organic forms become machinic, interlocking partial objects in an ecological economy. Forms depend on populations, populations imply codes, codes undergo decoding and displacement, natural selection does its work, and the rest is history.

However, this new account is further complicated by an internal inconsistency in the *Thousand Plateaus* theory of the organism. In the Tenth Plateau, Deleuze and Guattari quietly set aside the stratigraphic theory of organic form and with it the neo-Darwinism of their Geology of Morals, replacing it with an idea of symbiosis that is at odds with neo-Darwinism though increasingly favoured in philosophical biology.

## Life in the Strata

The idea that organic form 'depends on an autonomous code' means that species-identity is not a function of a whole ecology or even the whole organic stratum. The code is described as autonomous because it is supposedly as segregated from environmental interaction as Weismann's 'eternal' germ-line.[1] Today we call it the genome and equate it with the nucleic DNA sequence. This line, a linear string of organic symbols, moving across all the strata, is an expression, meaning a physical actuality, and a content, meaning a new quality. A class of molecules, nucleic acids, are the physical, molecular expression, and proteins the content, the emergent new quality. Nucleic acids express the content proteins, and, the authors say, this happens at the molecular level without any action from the exterior milieu. They say that coding, or morphogenesis, the evolution of genotypes on the organic stratum, 'takes place on an autonomous and independent line that detaches as much as possible from the second and third dimensions. . . . The essential thing is *the linearity of the nucleic sequence*' (Deleuze and Guattari 1987: 59–60).

That is probably a mistake. They may give undue credit to François Jacob's book, *The Logic of Life*, which they cite with approval many times, including the remark, 'In the living world, the order of order is linear' (1993: 286). Jacob advocated a genetic determinism now little favoured in biology. Like Deleuze and Guattari, he emphasised linearity. 'Evolution became possible through the relationship established between the structures of the organism in space and the linear sequence of the genetic message' (311). He also advanced an omni-competent genome. 'The whole plan of growth, the whole series of operations to be carried out, the order and site of the syntheses, and their coordination are all written down in the nucleic-acid message' (313). One used to hear this a lot in philosophical biology, but the tide has turned toward a less reductive account that understands whole organisms in whole ecologies to be indispensable to evolutionary change.

An autonomous genomic line may have had architectonic appeal for Deleuze and Guattari. It fits neatly into the theory of assemblage in the Geology of Morals. Through this one-dimensional line the whole organism in its every organ and function is exposed to the environment and natural selection. They say, 'The detachment of a pure line of expression on the organic stratum . . . enables [the organism] to put all its interior layers topologically in contact with the exterior' (1987: 59–60). Crystals grow from their surface. Organisms grow from their depth, and unlike the growth of crystals organic reproduction is highly sensitive to ecological change.

Nevertheless, a genome is not a single line. There is no one continuous genetic line for the organism or its organic form. 'The genome of virtually all metazoan cell lines consists not only of nuclear genes but of the genetic heritage of an often bewilderingly complex suite of endosymbiotes. The genome of an organism does not have to be concentrated in one spot within the organism, and it rarely is' (Peacock 2010: 235). Deleuze and Guattari's theory of multiplicities might have been more cogent (albeit more complicated) were they to have attenuated Jacob's neo-Darwinism and done more to integrate the thought on symbiosis that is prominent elsewhere in *A Thousand Plateaus*.

They also follow neo-Darwinism in assuming that the pure line of the genome is a complete program for the organism. Why else would they think that this line ensures that the totality of the organism is in topological contact with the environment? This again is something that we used to hear a lot, but less and less now from biologists. The deepest fault of the metaphor of DNA as a program of information is the very one Deleuze and Guattari like best, namely, that it implies a dualist framework of matter and information. A process that is intrinsically dynamic, embodied and environmentally embedded is projected into the reified abstraction of a generic line (Thompson 2007: 186).

Susan Oyama has been a leader in taking biological thought in a new direction. 'The environment . . . must be seen not only as a source of phenotypic variance, but as fundamental to species character, and not only as passive support, but as equal partner with the genes in giving rise to living beings' (2000: 29; cf. Robert 2004). What passes across generations are the resources needed for development, which include genes, but also cellular machinery, larger developmental complexes, the material reproductive system, even parental care – an entire evolved, species-specific developmental system. These developmental systems are an ecologically embedded network of many, typically self-organising biological individuals, of which the autopoietic cell is the paradigm (Thompson 2007: 206). The unit of evolution is this entire developmental system. Nothing more abstract will do. It would be impossible to equate developmental systems with a linear sequence of anything. 'Once genes are put back into the dynamic context of the cell, the picture of evolution and development that emerges . . . is one of a continuous and nonlinear causal spiral of interdependent factors at multiple levels of the life cycle, of which the genetic is only one' (185).

*Difference and Repetition* raised the embryo to a cosmic model. 'The entire world is an egg' (Deleuze 1994: 216). In *A Thousand Plateaus* the favoured model is the symbiotic assemblage of species; for instance, flowers and bees, orchids and wasps, host and parasitic tick, hunter and prey. We are only beginning to appreciate the ubiquity of symbiosis in the evolution of life. Every organ of every organism is a symbiotic assemblage of formerly dissociated organisms, captured by a vital centre of power and obliged to adapt. The new model is not perfectly adapted to the theory, however, and introduces tension into Deleuze and Guattari's account. When they are thinking of organisms stratigraphically (organic form), they tend toward a neo-Darwinism after Jacob's model. But when they are thinking about *natura naturans*, or physical becoming, the baffling complexity of their Geology of Morals recedes, and with it the neo-Darwinism, replaced by an idea of symbiosis that is at odds with neo-Darwinism, but increasingly favoured in philosophical biology. I will return to that. First, I want to pause over the idea of symbiosis, and some of the questions it raises for the philosophy of biology.

## Symbiosis

The basic idea of symbiosis is an assemblage of heterogeneous organic forms that persists for a long period relative to the generation times of the interacting organisms, and which typically leads to the emergence of novel metabolic capabilities in at least one of the partners. Variations on this theme range from parasitism to mutualism and from transient to permanent interdependence. Sometimes the symbionts are cyclical, acquired independently in each generation from the environment or acquired from parents soon after birth. Often they are inherited (Sapp 2004: 1046–56).

The phenomenon was well known in antiquity. Herodotus told of the Egyptian plover supposedly feeding on leeches from the mouth of crocodiles. With enough repetition it eventually became a fact. Aristotle repeated it in three different treatises, also reporting a more likely mutualism of the bivalve *pinna* and the crustacean *pinnotheres*. There were other accounts in Cicero and Pliny. They were accepted as evidence of nature's wisdom and balance or, in Christian authors, proofs of divine providence (Sapp 1994: 18–19).

Modern natural-historical authorities in the nineteenth century discounted these anecdotes and dismissed the notion of 'mutual assistance' as a naive anthropomorphism. What historian of biology Jan Sapp calls 'polite biological society' did not tolerate talk of symbiosis (2004). Partly that was because many of the first modern claims for symbiosis involved bacteria. The idea that bacteria could play a beneficial role in the tissues of plants and animals conflicted with the germ theory of disease. Pasteur inadvertently promoted the dogma of the asepsis of healthy animal tissue. To find vile microbes in healthy organs was offensive. Microorganisms were germs, carriers of disease. They cannot cooperate with life, only decay it.

Untrue, deeply untrue, but firmly held. Another reason symbiosis was side-lined was its clash with the central dogma of the Modern Synthesis to the effect that chromosomal genes were the sole basis of heredity. The so-called New Synthesis in modern biology was built on the premise that natural selection acted on small hereditary differences among individuals in interbreeding populations. Evolution was predicated on conflict. A hypothetical cooperation between individuals of different species seemed to many like an obvious dead-end and it was left unexplored. Only after about 1970 did the full scale of the oversight begin to come into view. Passive-aggressive suppression of symbiosis was finally lifted almost uniquely because of the work of one woman, Lynn Margulis (1993). She confirmed that cellular mitochondria are formerly free-living bacteria that symbiotically merged about two billion years ago.

Ever since this breakthrough, the most stimulating evidence of mutualism comes from the pervasive role of bacterial symbiosis in practically all life. Microbial symbionts carry out many chemical reactions otherwise impossible for their hosts to perform. Collectively they photosynthesise, fix nitrogen, metabolise sulphur, synthesise amino acids, provide vitamins and growth factors, and ward off pathogens. A species of worm lives in the gut of lobsters and eats only dead eggs whose decomposition would become toxic. Virtually all mammalian and insect herbivores would starve without cellulose-digesting bacterial symbionts. The termite gut contains thirty species of protozoa and two hundred kinds of bacteria. Collectively these symbionts weigh nearly as much as the termite itself. The greater part of land plants are mycorrhizal symbionts with soil bacteria. Symbiosis has even more intimately penetrated the interior of plants. Chloroplasts, the energy-generating organelles of all plants, derive from formerly free-living cyanobacteria. Photosynthesis is entirely bacterial. No alga or plant evolved photosynthesis on its own. 'All shared some ancestor – recent or remote – that ate but failed to digest a green or red or greenish blue bacterial photosynthesizer' (Margulis and Sagan 2002: 99).

The seemingly innocent assumption that symbionts must be in close proximity with one another obscures the true pervasiveness of symbiotic relations. It has been suggested that birds of prey are in a symbiotic relationship with the burrowing mammals they feed on, as are whales and schools of krill.[2] The lifecycle of these predators has become entrained with that of their prey, even though the prey are in direct physical contact with the predators only when they are eaten. Symbiotic cooperation has linked the evolution of species of baboons and Impala antelope in east Africa. Baboons follow the herds; their approach is scented by the antelope and welcomed. Baboons occupy the tree heights, from where they can see the distant approach of predators, while the antelope, with exceptional olfactory sense, are alert to dangers in the tall grass. The cooperation has evolved, the evidence being that each species innately understands the alarm calls of the other, taking countermeasures appropriate to the detected threat. As for human beings, Margulis estimates that 90 per cent of the $10^{14}$ cells in our body are bacteria of several hundred species. An adult human being is about ten per cent dry weight bacterial symbionts (Margulis and Sagan 2002: 18).

Symbiosis only seems unlikely if one is dogmatic about the omnicompetence of genes. One then expects the genetic line to be autonomous, abstract and isolated from environmental interaction except for natural selection's thumbs up or down for phenotypes. This is the classic Weismannian Doctrine that we see in Deleuze and Guattari's idea of form on the organic strata. As I mentioned, few in biology will now defend these pure abstract lines.

The question of where the variations come from among which natural selection selects has been a problem in Darwinism ever since Darwin. Symbiosis was unsurprisingly ignored as a potential source of genetic variety. The architects of the New Synthesis thought they had a better answer in advancing genetic mutation as the principal source of variety. However, as Margulis observes, the accumulation of mutations has never been shown, in the laboratory or in the field, to lead to a crossing of the species barrier. She finds mutation wildly overemphasised as a source of hereditary variation, and advances the hypothesis that symbiosis is not just a source but the principal source of variety in organic forms (Margulis and Sagan 2002: 11, 72). Symbiosis also seems to play a role in the generation of functional novelty. It may be part of the explanation both of rapid bursts in evolution, and the very existence of certain types of organisms. Cooperation can produce full-blown novel functionality instantaneously. When the new functionality confers survival advantage on cooperators and something of this tendency to cooperate is heritable, then it can be reinforced by natural selection, and become an evolved symbiotic cooperation (Peacock 2010: 248).

Margulis advances a strong thesis on symbiosis in the evolution of life. According to her 1997 view, it is 'increasingly certain that all visible organisms evolved though symbiosis', and that 'all species which have ever evolved have co-evolved' (Margulis and Sagan 1997: 33, 189–90). Five years later she is more adamant: 'All organisms that can be assigned to a unique species are products of symbiogenesis', all of the larger, more familiar organisms 'originated by symbiont integration that led to permanent associations', and 'all nucleated organisms are composite. All are products of integrated symbionts' (Margulis and Sagan 2002: 6, 142, 51). 'No life on earth consists of unassociated individuals of the same population.'

Symbiosis forces us to broaden our notion of what is heritable. Some symbiotic associations are themselves heritable since the genomes of the symbionts are passed on (usually maternally) to the offspring; it is not only nuclear genes that are inherited. Symbiosis also raises the old question of what a biological individual is. What constitutes biological autonomy? The more pervasive the symbiosis, the more tenuous the delimitation of an individual. Symbiosis signifies that protists, plants and animals are not single genetic entities. Perhaps the concept of organism needs to be defined in terms of a functional field that includes microbial communities (Sapp 2004). Biologist Joan Roughgarden observes that 'without exactly saying so anywhere, biologists define an individual as an entity containing one genome in one body. The problem turns up in species where one genome gives rise to multiple bodies [poplar trees, strawberries, beach grass], or where multiple genomes reside within a single body [lichens, corals, termite or bovine gut]' (2009: 8).

One genome in one body – that was also the assumption in the Geology of Morals' account of organic form. When in the Tenth Plateau *becoming* becomes the theme, Deleuze and Guattari's own assumption is in effect suspended in favour of a ubiquitous symbiosis whose abominable participations mock the pure line that supposedly generates organic form. What makes individuals individual is not the purity of their lines but the endurance of their symbiosis.

An alternative concept of biological autonomy or individuality takes for its model the single cell, in particular, the autopoietic reaction system that creates the membrane which in turn shelters cellular reactions. A biological individual is this autopoietic process in cells and in everything composed of cells, all life. A biological individual is a dynamic topological pattern in the changing material of cells and cellular systems, produced and maintained autopoietically by the system itself. The membrane is not a mere containment device for the reaction network, but is produced and maintained as a product of those reactions, and it is that which endows it with some degree of autonomy or individuality (Thompson 2007: 75, 105; cf. Varela 1979).

Finally, since our context is Deleuze, and since Deleuze is our best reader of Bergson, I should note that Bergson was among the majority sceptical of symbiosis or claims of its importance in the early twentieth century. For Bergson, evolution never associates but only dissociates, never converges but only diverges. Seeming examples of symbiosis (which he discussed under the older name of polyzoism), such as soil microbes and plants, are not, he says, 'produced by a reciprocal adaptation'. What looks like symbiosis is really is a trace of past unity rather than a creative association. He dismisses it as 'an exceptional and abnormal fact' incidental to life's evolution (1998: 117, 260). He said the same thing about bacteria, probably his biggest mistake. Evolution is dominated by microorganisms, which have always been the dominant form of life on earth, and as Margulis concluded, 'The speed, volume, and antiquity of bacterial gene-trading activities underlie the evolution of all the rest of life on Earth' (Margulis and Sagan 2002: 85).

## Abominable Participations

Jacob compared 'exceptional processes' like symbiosis to 'the "abominable couplings" dear to antiquity and the Middle Ages'. He says that 'were such encounters able, even exceptionally, to have consequences, this is enough to provide an opportunity for very profound changes' (1993: 311, 312). In other words, symbiosis could be a potent source of variation for natural selection to work on, if only symbiosis was not rare to the point of being almost a fable, as Jacob thought it was. Margulis recovers the same image and predictably goes much further: 'Attraction, merger, fusion, incorporation, cohabitation, recombination – both permanent and cyclical – and other forbidden couplings, are the main source of Darwin's missing variation' (Margulis and Sagan 2002: 205).

Scholars remain uncertain as to exactly what religious authorities condemned with the language of unnatural nuptials (Bullough 1982: 55–71). An early variation on the theme appears in Leviticus: 'Thou shalt not lie with mankind, as with womankind: it is abomination' (18:22, KJV). Paul promoted the terminology to connote otherwise unnameable sexual offenses. God wanted to smite the idolatrous pagans, whose irrepressible polytheism worshiped the creature more than the creator. 'For this cause

God gave them up into vile affections: for even their women did change the natural use into that which is against nature, and likewise also the man, leaving the natural use of the woman' (Rom. 1:24–7). The *Apostolic Constitutions* (third century) refer to 'unlawful mixtures', 'contrary to nature', 'wicked and impious'. Augustine explained the unnatural use of an organ as any for which it was not given by God, but which is imposed upon it by man's intention. Then he said that for a man to use a female organ in a way that is 'against nature' was 'abominable in a prostitute, but more abominable in a wife' (Bullough 1982: 59). By the twelfth century 'what is done against nature' had become an all-purpose euphemism for forbidden sexual activity. Such vagueness was not without its uses, but it made the terminology unsuited for the Handbooks of Penance, which, written to regulate the hearing of confession, had to be precise and discerning (McNeill and Gamer 1990).

Deleuze and Guattari seem to enjoy subverting the religious trope, describing symbiotic linkages across scales and kinds as unnatural nuptials, unnatural couplings and abominable participations. They say 'the plane of composition, the plane of Nature, is precisely for participations of this kind, and continually makes and unmakes their assemblages, employing every artifice' (1987: 258). There is really nothing unnatural about it. It's the way nature works. These mixings are unnatural only in judgements of God about what goes with what, as if speciation were a divine order that must be maintained. It may also be the way philosophy works, driven to conceptual invention by the desire to render consistent what is already 'inside' philosophy with something 'outside', different in form and substance, an alien species, as it were, like Greek medicine, geometry, Christianity, Galileo's telescope, or Boyle's corpuscles, which become new symbionts with philosophy, what Deleuze and Guattari call 'the not-external outside and the not-internal inside' of thought (1994: 59–60).

Their appreciation of the ubiquity of symbiosis in the evolution of life mostly unfolds in the Tenth Plateau, 'Becoming Intense, Becoming Animal, Becoming Imperceptible', and especially in their discussion of packs. Packs proliferate by contagion, not heredity. Bands proliferate by infection, like an epidemic, or by catastrophe, as on a battlefield, rather than by filiation or natural sexual reproduction. These promiscuous alternative ways of mixing – infection, contagion, invasion – link heterogeneous terms, like wasp and orchid, human and virus, Spanish and Native, and so on. The combinations are neither genetic nor structural but are rather those unnatural nuptials of symbiosis. The discussion of pack phenomena in this context is well motivated. The becoming of any enduring assemblage is a pack phenomenon. In biology, the developmental systems that are the most likely units of evolutionary change are precisely such packs. Their elements are heterogeneous, not belonging to one family, yet entrained, their changes synchronised. Their ensemble movement is a new quality on the earth, which propagates by expansion, occupation or peopling. Hence, as Deleuze and Guattari write, every animal, even every multicellular organism, 'is fundamentally a band, a pack' (1987: 239).

In this context they offer their most far-reaching claim for symbiosis. 'Unnatural participations or nuptials are the true Nature spanning the kingdoms of nature' (241). In these abominations we see 'the only way Nature operates – against itself. . . . The Universe does not function by filiation' (242). Symbiosis makes obvious the assemblage-character of every organism. 'An animal is defined less by its genus, its species, its organs, and its functions, than by the assemblages into which it enters' (Deleuze and Parnet 1987: 69). That is, its mixing or participation. On the one hand, assemblages are machinic. They are heterogeneous and collective, many operating together, bringing into play 'populations, multiplicities, territories, becomings, affects, events' (51). On the other hand, this operating or working together is not always cooperation, literally living together, which is what biological thought intends with the concept of symbiosis. Thus for example Deleuze says, 'Structures are linked to conditions of homogeneity, but assemblages are not. The assemblage is co-functioning, it is "sympathy", symbiosis' (52). That equates symbiosis and assemblage, which implies nothing biotic about symbiosis. They reiterate the point in another formulation, 'Each multiplicity is symbiotic; its becoming ties together animals, plants, microorganisms, mad particles, a whole galaxy' (1987: 250)

I doubt that this usage, equating multiplicity or machinic assemblage with symbiotic association, is a good model to follow. The idea of symbiosis is evolved cooperation among two or more former biological individuals. To make an evolutionary difference, the cooperation has to be heritable and reproduce in the species. It seems to me unsound to describe the conatus or endurance of any multiplicity or assemblage as 'symbiotic'. Machinic is a greater category. Not all machines are organic or symbiotic. The sugar that dissolves in Bergson's glass has mixed with the water and become an element of a new assemblage, but there is no symbiosis.

Deleuze defined as 'machinic' a system of relatively proximate things working together, subordinated to one centre of force. He explained that machinic 'does not mean either mechanical or organic'. It is their genus.

Machinic phenomena are a 'proximity grouping between independent and heterogeneous terms. . . . a proximity grouping of [e.g.,] man-tool-animal-thing. It is primary in relation to them, since it is the abstract line that crosses them and makes them work together' (Deleuze and Parnet 1987: 104). What holds machinic elements entrained, extracts collaboration and staves off flight or deterritorialisation, is the abstract machinic line that any given assemblage implements.

I think there is a point to resisting the monism that would make each assemblage or multiplicity subject to one conatus or power centre. That was Leibniz's thought for the monads, and it seems motivated to ensure the monarchy of God, but for the same reason the assumption is antagonistic to pluralism. A more consistent pluralism would acknowledge that the cooperation, be it unconscious and purposeless, of multiple vital centres can create effects that swamp single-centres on their own. A symbiosis is such an assemblage. It is a multiplicity, but multiplicity extended to centres of power too. The symbiosis expresses the cooperation of multiple biologically distinct centres of vitality. Not every multiplicity is biotic, but (if Margulis is right) practically any biotic assemblage is symbiotic, assembling many organisms, many species, an orgy of abominable participations.

Deleuze and Guattari seem not to appreciate what is singular about symbiosis, as contrasted with a mere machinic idea of parts. They overlook the creative cooperation of multiple centres of biological individuality without incorporation into a dominant master and higher-level individual. They apply the same analysis to [horse – rider – stirrup] as to [termite – gut bacteria] or [flowers – bees]. Human beings and tools – horse, rider and stirrup, say – form a machinic assemblage but not a symbiosis. The artifacts do not have a vitality in which the organisms cooperate. Tools would not exist apart from our need, and we and those needs would not exist apart from tools. But that is a mutual condition of existence, not a symbiosis. There are not multiple centres of life. There is one centre standing in relations of mutual coexistence with diverse things including tools. The conclusion I draw from the mutual coexistence of people and tools is that tools and users are abstractions of a symbiosis and ecological adaptation in which the categorical distinction possible for Aristotle between artifact and organism is not available. Which is what Deleuze and Guattari say too in at least one place. 'Tools exist only in relation to the interminglings they make possible or that make them possible.' Tools 'are inseparable from symbioses or amalgamations defining a Nature-Society machinic assemblage' (1987: 90). To be inseparable from symbioses is not to be symbiotic, as tools show.

There is a precedent for describing as symbiotic associated elements that are not organic forms. Freeman Dyson proposed that living cells, the simplest life, were products of a 'symbiotic' merger of two independently evolved prior systems. One of these was a self-replicating system of nucleic acids, the other a self-maintaining autocatalytic system of proteins (Dyson 1993). Margulis criticised the proposal on the ground that two inorganic systems cannot become symbiotic. Either they were both separately already alive, which would destroy the explanation, or first life emerges from their synthesis, which must then be a first cell, a single cell, and not a symbiosis with multiple centres of biological individuality. The crucial step for biological individuality is that reaction systems be enclosed within a membrane they produce. Prior to that, neither subsystem is alive, while after it is the whole system that is alive, which as Margulis said, makes the description of 'symbiotic merger' inappropriate (Margulis and Sagan 1995: 60).

Why should multiple, cooperating centres of life be important? Tools and *sapiens* enjoy mutually conditioned existence. *Sapiens* and our gut bacteria are multiple centres of reciprocal vitality living through each other in a way that is not parasitical but on the contrary mutually enhancing. The tools repeat mechanically. They have always already done all they can do. Under symbiosis, the other centres of life are alive and not mechanical. Tools remain as they are unless we change them. Symbionts struggle together, cooperatively, to become all that they can be. When I enrol a tool, I become more effective at what I already am. When I form a symbiotic relation with a second centre of life I enhance my potential, acquiring qualitatively new tendencies.

Cooperation with a second centre enhances what I can do, as it also does for the other reciprocally. Unfortunately, however, we cannot overlook parasitism, which is a mode of symbiosis, an extreme on the symbiotic continuum. Symbiosis is not invariably a mutually enhancing cooperation. Parasitism is joyous only for one part of the symbiotic assemblage. The host is reduced to bare life, captured and made a resource, compelled to live as a mere resource, a proper slave, Aristotle's living instrument (*Politics*, 1253b). Reflection on parasites should remind us that every symbiosis is an adaptation as blindly entered into and as unconsciously effected as any other, and invariably has some degree of latent hostility.

## Anorganic Life

Deleuze and Guattari refer to a tendency or line of flight that cannot be confined to any stratum, an unmitigated power of absolute deterritorialisation that they describe as anorganic (sometimes non-organic) life. For instance, the matter-flow of metals is 'non-organic life'; metal 'is neither a thing nor an organism but a body without organs' (1987: 411). I do not think they want to say that the flowing metal is alive, or that it is a case of *in*organic life. The flow of metals is *like* vitality in being a flow rather than a form. The difference is that metal does not flow at all (except deep in the earth) without human artifice, and the forms it takes are never autopoietic individuals. The forms are endlessly changing but not evolving. The metal is always already everything that it can be. Even forms no smith has invented are in the present competence of the metal, unlike a biological individual, which is never all that it can be until it is no more.

This anorganic life is, therefore, not a panpsychic vitalism. It does not draw artificial and inorganic things into the circle of life. Anorganic life is a power of tendency without organisation or organs. It is all becoming and nothing formed. 'The life in question is inorganic, germinal, and intensive, a powerful life without organs, a Body that is all the more alive for having no organs' (499). Anorganic life is not a species of life. It is a becoming, becoming species and becoming individual, but this becoming is not itself any accomplished act. It is a power, a tendency, not an individual with a form, and it is a duration, an interval still open, an unfinished event. 'Life tears itself free from the organic by a permutating, stationary whirlwind' (499).

Bergson called this anorganic life *élan vital*. In the terms of *A Thousand Plateaus*, anorganic life is a power of apparently unconditional deterritorialisation, a line of flight that escapes every stratification, cutting through everything assembled, leaving every vital centre behind like a trace or footprint in the sands of time. 'What movement', they ask, 'what impulse, sweeps us outside the strata?' Whence the power of deterritorialisation? It is life, a vitality that is not limited to the lines that supposedly define organic forms on the organic strata. It is therefore described as anorganic, meaning the vitality or vital becoming that leaves organic forms behind it as it breaks into a trackless future. 'Not all Life', Deleuze and Guattari write, 'is confined to the organic strata: rather, the organism is that which life sets against itself in order to limit itself, and there is a life all the more intense, all the more powerful for being anorganic' (503). Of course, there would be no thought of enclosing life on the organic strata were it not for the neo-Darwinism they imported into their Geology of Morals. From the point of view of their Tenth Plateau and from the perspective of biological symbiosis, the becoming of species is already an affair not of one special stratum of organisms but, as they said, 'a whole galaxy'.

They identify this anorganic vitality with the power of deterritorialisation. The line that carries away is behind every organic form – or rather, out ahead, the form being a trace of a power that jets into the future. The connection they make between deterritorialisation and anorganic life seems to be an implication of their Bergsonism. Deterritorialisation is a tendency, a vast tendency, the vastest, entropically drawing all things apart. Bergson outright equated *élan vital* with tendency – 'Life is tendency' (1998: 99). He did not mean life was *a* tendency, one among many. He says it *is* tendency. That means the origin of tendency. Matter alone has no tendency. A molecule, a crystal, or a stone has always already been all that it can be, and changes only when moved by external force. Organisms, no matter how simple, *tend*, and do not mechanically repeat the past. To describe life as tendency is not to ascribe it plan, purpose, design or end. The idea of tendency is movement in a particular direction without prior design. A tendency is tending, leaning, verging, *becoming*. It *is* (actually) nothing at all.

Life introduces this tending and becoming into material systems that would not have it on their own, and the result is tendencies that are not naturally harmonious. Over time some have become more powerful than others. Some were lucky or had superior allies. The balance of tendencies in nature and especially on the earth is intrinsically unstable and constantly changing. There is tension, a contest, a struggle for actuality or materialisation. It is the struggle for presence, for becoming present. The tendency of deterritorialisation is the tendency of life, of change, of becoming, of duration, of time itself. This is perhaps Deleuze and Guattari's deepest extension of Bergson's thought, to use the new concept of deterritorialisation to elucidate the connection he implied between time and life. *Élan vital* is not just the becoming of life; it is the being of time.

Anorganic life is their word for *élan vital*, and *élan vital* is another word for becoming, for endurance, for duration, tendency, even time itself. The quality of time that Bergson stressed was interpenetration. Time has no parts. The least degree of change anywhere changes everything everywhere (which takes time). So the least change in any environmental factor (deterritorialisation) must elicit some change at some scale from every organism. You cannot alter an ecology without changing the *tendency* of the whole.

Deleuze revealed what he frankly described as his 'profession of faith' as an empiricist. He made an unlikely reference to the little-known American poet-philosopher Benjamin Paul Blood and his work, *The Anesthetic Revelation and the Gist of Philosophy* (1874). 'Nature is miracle all. She knows no laws; the same returns not, save to bring the different. The slow round of the engraver's lathe gains but the breadth of a hair, but the difference is distributed back over the whole curve, never an instant true, ever not quite.'[3] This 'ever not quite' delighted Deleuze (as it did William James). The engraver's lathe was an original image of what C. S. Peirce described as an infinitesimal tendency to diversification, and Bergson's constant creation of the new.[4] Every movement is new and has to be redistributed over the whole past, infinitesimally altering every trajectory.

## Ironic Atomism

Not just any part in any machine is a symbiosis. The parts have to be alive, they have to be living parts of the larger life of the assemblage, as they usually are not, except in an organism, which is therefore a special sort of machine. Leibniz wrote of organic machines. Only God can make them. Bergson more plausibly thought that 'machine' is an anthropocentric perspective useful for prediction and control but never true to nature, which is in no way mechanical (1998: 226). The being of nature is the being of becoming, and becoming cannot be mechanical; a simple repetition of the past becomes nothing. A machine is a way of staving off becoming, slowing it down, modulating it, synchronising its changes to create relative stability across multiple centres.

However, becoming could be described as machinic, inasmuch as what becomes are new assemblages, though the becoming is not a mechanical transition, an instantaneous flip-flopping from one mechanical state to another. There is no repetition. Becoming is duration, and duration is, in Bergson's term, pure heterogeneity. 'Pure duration might well be nothing but a succession of qualitative changes, which melt into and permeate one another, without precise outlines, without any tendency to externalize themselves in relation to one another, without any affiliation with number . . . pure heterogeneity' (2001: 104). He means that any two moments of duration are qualitatively different regardless of the interval between them. Let the difference of time be as small as you like, the two moments will be qualitatively different. Time refuses to be cooped up in a changeless present, however brief. The more one tries to make time stop, the faster it divides and changes. Not even Achilles can catch up.

Becoming, duration, anorganic life, *natura naturans* and deterritorialisation are so many designations of an uninterrupted, unfinished movement dis-aggregating and re-aggregating assemblages, much as Epicurus proposed, for whom what *becomes* are molecules, compound bodies. Atoms do not become anything. They are always already everything they can be. It has been well said that 'the fundamental category of matter in Epicurean physics is "body" and not "atom"' (Morel 2009: 79). Are there atoms in the Deleuzian philosophy of nature? In *Difference and Repetition* he called them pre-individuals. These ultimata are not actual atoms, however, but virtual haecceities, or qualitative singularities. If we ask what in Deleuze's philosophy of nature is not machinic, not a multiplicity or assemblage, the answer seems to be any haecceity, any virtuality, any quality. Machinic is the mode of actualisation in a world of original plurality. What is not actual – what is, for instance, past or virtual – is not assembled, not an assemblage. There are no unnatural couplings in the virtual because there are no couplings, no inclusions and exclusions, no cellular membranes, no autonomy or individuality. Everything virtual abides with everything else. But to become, to actualise or materialise, requires incompossible tendencies to struggle, find allies and contract symbioses.

Actualisation is a will to power. This is no mere homage to Nietzsche. A 'will to power' is the tendency to struggle, the tendency to enter time, to submit to contingency, and become all that a tendency can be. The 'will to power' is not a will to dominate others; it is an agonistic will to contest and struggle in action with multiple centres of vitality.[5] What is willed is the contest, but the domination of a single centre of power is not the only possible outcome. This 'will to power' of the virtual is its tendency to do all that it can to actually, historically become all that it virtually is. To be virtual is to be ever tending to be actual, tending to leave the virtual condition where everything is compossible, and actualise selectively and competitively. Part of what it means to describe existence as virtual is to attribute this tendency and relative power to actuality, a tendency of all virtual being to do all that it can to actually become all that it virtually is. When that tendency is exhausted, the individual becomes a memory.

Nothing actual is without tendency. Actuality is tendency pressing, tending. All tendencies tend, but to pass the threshold and happen to be depends in part on what has already happened. Is there a fit? Are there openings, voids? Actual environments perform a non-teleological selection on tendencies, presenting to some openings to actuality and closing off others. Atoms require voids. An atomism of the virtual, where the timeless atoms are virtual haecceities, requires voids of actuality, voids in the actual where something virtual and tending can become present. Actualisation cannot be completely conditioned by the past. There must be voids. The great void is the future. It is all void. The past is full, the future void. What will pass across the interface and become the future? The past does not determine that completely, because the past is not finished. The past is ceaselessly changing, ever not quite. That is the reality of the future – that the past is changing and therefore cannot completely, rigidly determine the future. If the past were not changing, then the future would not exist. Time would not exist.

We cannot be sure what future will fit with the whole past until it happens. Each present slightly modifies every trajectory of the past. We imagined straight what was imperceptibly curving. The future is the void that lets virtual atoms (pre-individual haecceities) form actual molecules (assemblages). The atoms are virtual tendencies. The void is the future. It is void because the past is not a straight line. It is infinitely curving, infinitely differentiating, a line of pure heterogeneity void at every present. It is never clearly and distinctly determined what 'the same' would be *now*.

Qualities are the content of duration, and haecceities are virtual qualities, obscure but distinct tendencies manifesting as differences among actual qualities. The result is an ironic atomism, whose particles invert those of the Epicurean atom, being indistinct and never actual, enjoying the timeless being of the virtual, more like the gods of Epicurus than his atoms. From Democritus to Dalton, atoms have been associated with empiricism. Deleuze explains empiricism as a theory of concepts in the wild. 'The genius of empiricism, which is so poorly understood, [is] the creation of concepts in the wild, speaking in the name of a coherence which is not their own' (2004: 141–2). He means, I think, that concepts have to be invented – a classical empiricist refutation of 'innate ideas' – and their invention is a response to problems that refer to an evolved form of life. If concepts have to be invented from perceptual material and tested in the experiments of life, the same goes for the evolution of biological species. Species are assemblages, symbiotic in the proper sense of multiple centres of contingently evolved life. The very existence of such species is experimental. In evolution as in individual life, it is not possible to say in advance what can symbiotically associate with what. The experiment has to be tried.

In *Difference and Repetition*, Darwin is the thinker of the individual, fruitfully reversing the priority classical thought accords the generic. By the time of *A Thousand Plateaus*, Deleuze and Guattari take on a rather doctrinaire genetic neo-Darwinism, the most interesting feature of which is the way they inconspicuously contradict it by developing a theory of evolutionary symbiosis that undermines the supposed omnipotence of the genome. The result restores and fruitfully complicates the earlier theory of the individual, now understood as a symbiont of symbionts. Symbiosis cannot be reduced to machinic assemblage. Life introduces something new, specifically, virtuality. The anorganic life they identify with absolute deterritorialisation is Bergson's *élan vital* updated for an ontology of virtual tendencies animating symbiotically evolved assemblages.

## Notes

1. On Weismann's germ-line and the important argument that this is a product of evolution and not a formal parameter of evolution as such, see Buss 1987.
2. The examples of predator birds and whales are from Peacock 2010, and the baboons and antelope from Dröscher 1965: 93.
3. Benjamin Paul Blood, in Deleuze 1994: 57. Deleuze cites Blood from Wahl 1925, who cites from James 1987: 1312.
4. James compares Bergson and Peirce on this point in 1987: 815–16.
5. This is admittedly against the grain of Deleuze 2006. The explanation of will to power postulates that every force is related to other forces uniquely in terms of domination-submission (chap. 2).

## References

Bergson, H. (1998), *Creative Evolution*, trans. A. Mitchell, Mineola: Dover.

Bergson, H. (2001), *Time and Free Will*, trans. F. L. Pogson, Mineola: Dover.

Bullough, V. L. (1982), 'The Sin Against Nature and Homosexuality', *Sexual Practices and the Medieval Church*, ed. V. L. Bullough and J. Brundage, Buffalo: Prometheus, pp. 55–71.

Buss, L. W. (1987), *The Evolution of Individuality*, Princeton: Princeton University Press.

Deleuze, G. (1994), *Difference and Repetition*, trans. P. Patton, New York: Columbia University Press.

Deleuze, G. (2004), *Desert Islands and Other Texts (1953–1974)*, ed. D. Lapoujade, trans. M. Taormina, Los Angeles: Semiotext(e) Foreign Agents Series.

Deleuze, G. (2006), *Nietzsche and Philosophy*, trans. H. Tomlinson, New York: Columbia University Press.

Deleuze, G. and F. Guattari (1987), *A Thousand Plateaus: Capitalism and Schizophrenia*, trans. B. Massumi, Minneapolis: University of Minnesota Press.

Deleuze, G. and F. Guattari (1994), *What is Philosophy?*, trans. H. Tomlinson and G. Burchell, New York: Columbia University Press.

Deleuze, G. and C. Parnet (1987), *Dialogues*, trans. H. Tomlinson and B. Habberjam, New York: Columbia University Press.

Dröscher, V. B. (1965), *The Mysterious Senses of Animals*, trans. E. Huggard, New York: E. P. Dutton.

Dyson, F. (1993), *Origins of Life*, Cambridge: Cambridge University Press.

Jacob, F. (1993), *The Logic of Life*, trans. B. E. Spillman, Princeton: Princeton University Press.

James, W. (1987), 'A Pluralistic Mystic', *Writings 1902–1910*, New York: Library of America.

McNeill, J. T. and H. M. Gamer (eds) (1990), *Mediaeval Handbooks of Penance*, New York: Columbia University Press.

Margulis, L. (1993), *Symbiosis in Cell Evolution*, 2nd edn, San Francisco: W. H. Freeman.

Margulis, L. and D. Sagan (1995), *What is Life?*, New York: Simon and Schuster.

Margulis, L. and D. Sagan (1997), *Microcosmos*, Berkeley: University of California Press.

Margulis, L. and D. Sagan (2002), *Acquiring Genomes*, New York: Basic Books.

Morel, P.-M. (2009), 'Epicurean Atomism', *The Cambridge Companion to Epicureanism*, ed. J. Warren, Cambridge: Cambridge University Press.

Oyama, S. (2000), *Evolution's Eye: A Systems View of the Biology-Culture Divide*, Durham, NC: Duke University Press.

Peacock, K. A. (2010), 'Symbiosis in Ecology and Evolution', *Handbook of the Philosophy of Science*, vol. 11: *Philosophy of Ecology*, ed. K. DeLaplante, B. Brown and K. A. Peacock, Amsterdam: Elsevier, pp. 235–66.

Robert, J. (2004), *Embryology, Epigenesis, and Evolution: Taking Development Seriously*, Cambridge: Cambridge University Press.

Roughgarden, J. (2009), *The Genial Gene: Deconstructing Darwinian Selfishness*, Berkeley: University of California Press.

Sapp, J. (1994), *Evolution by Association: A History of Symbiosis*, New York: Oxford University Press.

Sapp, J. (2004), 'The Dynamics of Symbiosis: An Historical Overview', *Canadian Journal of Botany* 82, pp. 1046–56.

Thompson, E. (2007), *Mind in Life: Biology, Phenomenology, and the Sciences of Mind*, Cambridge, MA: Harvard University Press.

Varela, F. (1979), *Principles of Biological Autonomy*, New York: Elsevier North Holland.

Wahl, J. (1925), *The Pluralist Philosophies of England and America*, trans. F. Rothwell, London: Open Court.

Chapter 2

# The Egg: Deleuze Between Darwin and Ruyer

*Jon Roffe*

Gilles Deleuze's discussion of Charles Darwin in *Difference and Repetition* presents an interpretive difficulty. The key remark is the following:

> Darwin's great novelty, perhaps, was that of inaugurating the thought of individual difference. The leitmotiv of *The Origin of Species* is: we do not know what individual difference is capable of! We do not know how far it can go, assuming that we add to it natural selection . . . The great taxonomic units – genera, families, orders and classes – no longer provide a means of understanding difference by relating it to such apparent conditions as resemblances, identities, analogies and determined oppositions. *On the contrary*, these taxonomic units are understood on the basis of such fundamental mechanisms of natural selection as difference and the differenciation of difference. (Deleuze 1994: 248)

The interpretive difficulty arises when we ask how the notion of 'individual difference' relates to the other major categories of Darwin's thought, above all those of the *species* and the *individual*.

According to the classical reading, the subject of both Darwin's account and of evolution itself are species. Now, when Darwin comes to consider the concept of species – and indeed, every one of his descriptive categories – it is similitude that arbitrates. A species is defined in terms of its trans-individual similarities. As Darwin writes at the very start of the first chapter of the *Origin*, 'When we look to the individuals of the same variety or sub-variety of our older cultivated plants and animals, one of the first points which strikes us, is, that they generally differ much more from each other, than do the individuals of any one species or variety in a state of nature' (1859: 10). And when it comes to the development of a species, the natural scientist's gaze is trained on the emergence of distinct convergent traits, such that: 'a well-marked variety may be justly called an incipient species' (52).[1] For Deleuze, categories of this kind – those associated with the notion of similitude – are evidence of a representational ontology in which difference is necessarily 'incarcerated' (1994: 50). Indeed, as he says, the expectation of similitude in perception (the perceptions of empirical science, for instance) is one of the ways in which individual difference is misrecognised and undermined in thought. Consequently, 'the subordination of difference to identity and difference to similitude must be overturned in the same movement' (212).

It is also significant that the term 'species' is absent from Deleuze's discussion of Darwin in *Difference and Repetition*, a concept that he subjects to a withering critique throughout the book. The Aristotelian origin of the category is first subject to a sustained and withering criticism from which it emerges as a kind of metaphysical error resulting from Aristotle's misprision of difference. By the final chapter, this error appears as a transcendental illusion. Species appear to be natural types because of the very nature of the intensive individuation of species themselves. This is to say that species come last in the order of ontogenetic process, but they appear to the habitual and habituated human thinker to be first. Deleuze insists on the serious nature of the inversion of this order: 'In fact any confusion between the two processes, any reduction of individuation to a limit or complication of differenciation, compromises the whole of the philosophy of difference' (247). In positive terms, 'It is not the individual which is an illusion in relation to the genius of the species, but the species which is an illusion – inevitable and well founded, it is true – in relation to the play of the individual and individuation' (250).

So, from the point of view of both Deleuze's critique of representation and his attack on the notion of species, species-central readings of Darwin are very clearly at odds with the project of *Difference and Repetition*. If we turn to those readings of Darwin that emphasise the organic individual instead, it quickly becomes apparent that they fare no better. On this view, the species level of analysis is strictly nominal, bearing no ontological significance and only acting as a shorthand for numerical collections of similar living beings.[2] The individual assumes its full importance as the site of mutation and thus the source of life's auto-aberrant transformation on this view, one close in a superficial sense to Deleuze's own view.

It is true that this sense of the species as an extensive multiplicity (in contrast to the intensive multiplicity of the pack championed in *A Thousand Plateaus*) cannot be used to explain anything for Deleuze.

It is equally true, though, that not any conception of the individual will suffice for a philosophy of difference. This reading of Darwin's *Origin* treats living beings in their biological and material particularity as the self-identical subject of selection: 'each organic being', 'each individual beetle', 'each Proteolepas' (1859: 84, 136, 148). But what Deleuze understands by 'individual' and 'individual difference' is derived from his reading of a strand of idealist philosophy beginning with Leibniz and is only accessible by way of transcendental reasoning. What his notion of individuality entails will be addressed in what follows; the least we must recognise immediately is that Deleuze's concept of the individual, and the methodological perspective it evinces, are incompatible with both Darwin's finite organic individual and his empirical naturalism.

In sum, then, Deleuze's characterisation of Darwin as a thinker of individual difference evidently runs counter to Darwin's own position, whether this is construed in terms of real existence of species or the ontological primacy of the organic individual. Given this, what sense can be given to the heavy use that Deleuze makes of biology in *Difference and Repetition*? To be more precise: given that Darwin is not the key reference point in the elaboration of a dynamic ontogenesis, who is? The answer, I contend, is Raymond Ruyer, a still little-known thinker in the battery of Deleuze's intellectual influences.

To justify this claim, I would like to examine the sequence of biologists that Deleuze discusses in *Difference and Repetition*, a sequence that appears right after his invocation of Darwin and which culminated with the invocation of Ruyer. The topic under examination throughout the sequence is that of *embryogenesis*, which appears as the figure that encompasses his entire argument about ontogenesis in the decisive final chapter of *Difference and Repetition* on the assymetrical synthesis of intensity. For Deleuze, to understand the egg is ultimately to understand the dynamic reality of being.[3] And Deleuze's way of elaborating this point is greatly illuminated by reading him by way of Ruyer.

## Deleuze with the Embryologists

As I have just said, the case of the egg enjoys a particular significance in the account of ontogenesis that Deleuze develops in the second half of *Difference and Repetition*. Indeed, it first appears as an immediate response to the question: 'How does actualisation occur in things themselves?' (1994: 214). Throughout the pages devoted to the topic, the broad import of the egg and of embryogenesis is only reinforced by the assertion – made not once, but three times – that 'The world is an egg.'[4] This claim is, however, far from an evocative metaphor. Deleuze is quite specific about the sense in which the egg is a figure for ontogenesis as such, linking it to the major moments of this process on his account: 'the egg, in effect, provides us with a model for the order of reasons: (organic and species related) differentiation-individuation-dramatisation-differenciation. We think that difference of intensity, as this is implicated in the egg, expresses first the differential relations or virtual matter to be organised' (251). These remarks appear as a part of a fleeting and yet far-reaching summary of the history of modern embryological thought. Prior to the early nineteenth century, accounts of the embryo tended, on the one hand, to recapitulate scholastic, Aristotelian views, for instance in the work of Johann Meckel, who argues that the development of the embryo proceeds along the great chain of being – an early version of Ernst Haekel's famous claim that 'ontogeny recapitulates phylogeny'. On the other hand they tended to advance along the lines of the early modern mechanism exemplified in Newtonian physics. This latter perspective presented the embryo as already composed of discrete units: the embryo is, from the start, a mosaic. In turn, this debate intersected with another concerning the nature of organic specificity – do particular beings gain their specific features epigenetically, or due to a pre-existent form? – whose major combatants were Cuvier and Geoffroy Saint-Hilaire.[5] But Deleuze's reconstruction of modern embryology begins by invoking one of Meckel's (and later Darwin's) most important early critics, Karl Ernst von Baer (1792–1876). In the terms of the second debate, von Baer advanced a modified epigenetic account of ontogenesis. He demonstrated that, at the early stages of embryogenesis, different kinds of animals share very general features – for instance, early in their development, both mammals and birds possess gill slits. However, as the embryo develops, this level of general homology is subject to increasing specification – or, as he puts in the second of his four laws of development, 'The less general forms develop from the most general forms, and so on, until finally the most specialised form arises' (Churchill 1991: 11). Based on the science of his day, and following Cuvier, von Baer then argued that there are four basic kinds of living beings: vertebrates, molluscs, articulates and radiates. These four kinds exhibit what von Baer calls the 'concrete purposiveness' exhibited in all life, but a purposiveness that is not a determinism, necessarily making room for the irreducible play of contextual factors in order to explain the possession of the specific characteristics exhibited by any given living being (Von Baer, cited in Vucinich 1988: 97).

While he later conflates the two positions, when Deleuze first presents von Baer's position, he takes issue with it and invokes as a corrective the work of Louis Vialleton (1859–1929).

> In the first place, the highest generalities put forward by Baer are generalities only for an adult observer who contemplates them from without. In themselves, they are lived by the individual-embryo in its field of individuation. Furthermore – as Vialleton, a disciple of Baer, points out – they can only be lived, and lived only by the individual-embryo: there are 'things' that only an embryo can do, movements that it alone can undertake or even withstand (for example, the anterior member of the tortoise undergoes a relative displacement of 180 degrees, while the neck involves the forward slippage of a variable number of proto-vertebrae). The destiny and achievement of the embryo is to live the unlivable, to sustain forced movements of a scope which would break any skeleton or tear ligaments. (1994: 214–15)[6]

This critique is animated by the same reasoning that, earlier in the book, Deleuze deploys against Aristotle: to posit natural kinds is ultimately to adopt the perspective of the being already become, the 'adult observer'. Von Baer's account is therefore, despite itself, no fundamental theory of epigenesis, or even of genesis at all, since all so-called genesis is necessarily a *fait accompli*. Vialleton's intervention is thus decisive for Deleuze, since it amounts to the assertion that an adequate thought of the egg is one from the point of view of the egg itself in its dynamisms, an embryogenetic perspective. Not only does this perspective give these dynamisms their own *ratio* and significance, it rules out any reference to fixed forms, and any return to Aristotle.

In the work of two later embryologists, Charles Manning Child (1868–1954) and Paul Alfred Weiss (1898–1989), however, Deleuze detects a reiteration of the same broad commitment to fixed distribution and structure over an essentially dynamic process that begins in a purely differential intensive field. The main contribution of Child, initially a zoologist who did pioneering work on reflexive behaviour in butterflies, is thought to be his account of the organisation of organic development (but also regeneration). He was the pioneer in arguing that developing tissues, at every level of organisation up to the full adult state, are structurally organised around a frontal-dorsal or anterior-posterior axis. Rather than being a physically discrete feature of the developing animal, something which clearly does not exist at all stages of development (for instance, in the first divided cells), this axis is a differential biochemical gradient. Adaptive responses to the environment take place in the portion of the developing being with the fastest metabolism, which determines the course of development in a cascading fashion (cf. Hyman 1957: 79). In sum, every organism at every moment of its development already possesses a structural organisation around which this development takes place. While Weiss was concerned less with embryogenesis, and more with the neighbouring case of tissue regeneration, his overriding theoretical commitments are entirely in keeping with Child's. Though he was a determined critic of purely causal explanations of embryogenesis, and despite his strong sense of the dynamic character of biological life – 'In contrast to a machine, the cell interior is heaving and churning all time' (Weiss 1973: 39), he writes – he remained committed to inherent structural constraints characteristic of biological processes including those belonging to the embryo.

Here is Deleuze:

> Individuating difference must be understood first within its field of individuation – not as belated, but as in some sense in the egg. Since the work of Child and Weiss, we recognise the axes or planes of symmetry within an egg. Here too, however, the positive element lies less in the elements of the given symmetry than in those which are missing. (1994: 250)

Structure and fixity, organisation and species do indeed exist, but these must be conceived of as the result of the process of intensive individuation already in play *in* the egg. Now, of course neither Child nor Weiss held to anything resembling preformationism or von Baer's generic biotypology, but in taking this structural symmetry to be a natural given they both cling to an ontologically primary sense of a fixed structure.

It is at this point that Deleuze invokes the work of Albert Dalcq (1893–1973). Indeed, not only does Dalcq's embryology appear at first blush to exemplify the dynamic account of ontogenesis that Deleuze himself advances, but Deleuze's most technical account of the development of the embryo makes prominent use of Dalcq's own analysis. Deleuze quickly identifies the major categories animating Dalcq's work:

> Individuation always governs actualisation: the organic parts are induced only on the basis of the gradients of their intensive environment; the types determined in their species only by virtue of the individuating intensity. Throughout, intensity is primary in relation to organic extensions and to species qualities. Notions such as 'morphogenetic potential', 'field-gradient-threshold' put forward by Dalcq, which essentially concern the relations of intensity as such, account for this complex ensemble. (1994: 251)

Putting aside the fact that all of these concepts are derived from other, earlier thinkers, some of whom Deleuze directly cites,[7] they do appear to name features of ontogenesis that Deleuze's own account of (intensive) individuation and differenciation describe: an intensive field, populated not with kinds or types, but divergent lines and thresholds of variation. Dalcq's book *L'oeuf et son dynamism organisateur* appears, duly, in the tabulated bibliography that Deleuze included in *Difference and Repetition* with the description: 'Intensity, individuation and biological differenciation' (336).

A first reading of *Difference and Repetition* might therefore take Dalcq to be the source of Deleuze's own particular enthusiasm for the case of the egg. But to turn to Dalcq's own work is to discover – as in the case of Darwin – a body of work very much at odds with both the letter and the spirit of Deleuze's. Determining whether this results, again, as in the case of Darwin, from a heterodox reading or from a misreading is beyond the scope of the current chapter.[8] What must be noted, in brief, is that Dalcq's work is ultimately – despite several other dalliances we will review shortly – subordinated to a biochemical paradigm, in which only direct causal relations hold.

When Deleuze cites Dalcq's admittedly excellent phrase 'nothing is *a priori* caudal' (251), then, he dramatically overstates the break with a priori thinking that Dalcq manages to accomplish, since for him the tail is in fact preceded by a field of cells that acts to bring about the tail, just as every limb is preceded by a 'limb field' (Waddington 1966: 112).[9] We are, with Dalcq, a long way from the 'more subtle theatre' of Deleuze's intensive field of individuation (1994: 254, quoting Leibniz). A franker judgement of Dalcq's significance for an ontology equal to the embryo, or at least a judgement in keeping with Deleuze's overall project in *Difference and Repetition*, would probably read more like this: 'From 1935 to 1947, he oscillated between the affirmation of a specific vital activity, a *de facto* finality, and the purely physicochemical explanation. The title of Dalcq's major work, *The Egg and Its Organizing Dynamism* [1941], is altogether misleading: only gradients of substance and chemico-differentiations are at issue in the text' (Ruyer 2016: 194).

## Ruyer Between Mechanism, Vitalism and Preformationism

These words were penned by the philosopher Raymond Ruyer, a figure that at first glance appears to be as marginal as Dalcq himself for the project of *Difference and Repetition*, only garnering – like Dalcq – a single invocation in the text. But in this case, the scarcity of citation significantly misleads. If, as Deleuze asserts, the entire world is to be conceived as an egg, then the significance of a reference to a philosopher whose central guiding phenomenon is embryogenesis, and whose final posthumously published work is titled *The Embryogenesis of the World and the Silent God*, is not to be trivialised out of hand (Ruyer 2013).[10] Moreover, given that Deleuze's own account of the embryo qua expressive individual is profoundly influenced by Leibniz, the fact that Deleuze, as he works through the problem of individuation, will invoke a thinker he later identifies as 'the most recent of Leibniz's great disciples' (1988: 137) is no small matter. Finally, and more concretely, as we will now see, Ruyer not only, like Deleuze, criticises the major contemporary alternatives for conceiving of the embryo, he also advances an account of embryogenesis that Deleuze will endorse in most of its details.

Given the limits of space, it will not be possible to explicate in detail the significance of Ruyer for Deleuze's project in *Difference and Repetition*. I will instead make three summary points concerning his proximity to this project.

## *Three critiques*

The first point concerns Ruyer's critical approach to the trajectory in embryology just outlined. Indeed, as he points out at the beginning of *La genèse des formes vivantes*, the entire problem turns around the supposition of a static *logos* (embry*ology*), when embryo*genesis* is a process that involves a dynamic transition through states whose structures differ completely from each other, while still being states of the *same being*. At issue is the possibility of a dynamic '*unitas multiplex*' (2016: 14).

Ruyer's remark on Dalcq in fact indicates the three alternatives found throughout the embryological literature up to that point: mechanism, vitalism and preformationism. He is throughout his work most severely critical of the first option, which he considers to be riddled with explanatory deficits. Mechanism – purely physical interactions operating *partes extra partes* between established entities, the simple 'combination of physical and chemical processes' (194) – cannot explain the identity-in-difference characteristic of embryogenesis, this 'augmentation in complexity' (Ruyer 1958: 90; cf. 2016: 55). But neither can it explain the equipotent character of embryonic cells, such that intervening in the embryo to cut and rearrange its parts can, to a significant extent and for a significant period, still produce a viable human being. This equipotentiality is even capable of overcoming the *absence* of major portions of the initial embryo. As the example of the sea urchin egg shows, each of the two cells produced by the first division, when separated, produce a complete adult sea urchin. The same fact can be seen in the case of the starfish and, even more dramatically, in that of the hydra: 'With the exception of the tentacles, any fragment of a hydra can, without absorbing any nutrients, reconstitute the missing parts by rearranging those that remain' (1958: 93).

The particular character of what Ruyer calls 'internal reproduction' – reproduction that does not proceed through the adjunction of new cells but through the splitting into two copies of the parent cell – also remains inexplicable on the mechanist model. Not only are these twinned cells incapable of being produced by the simple accretion of parts, but this accretion cannot explain how the subsequent cells can come to be a part of a single organism.[11]

The immediate alternative to this approach – preformationism – is to insist that there must be a kind of fixed pattern or form that governs the development of the embryo. This preformationist alternative is what Ruyer also calls *de facto* finalism – characterised as such because it is often an implicit recourse when the mechanist account is found lacking.[12] It is also, for Ruyer, a clear non-starter. It first of all cannot account for the dynamic reality of the embryo's *unitas multiplex*, which never possesses the self-same form from one moment to the next. Preformationism also presupposes a certain scale of application – the organism as a whole – and thereby once more overlooks internal reproduction. The question 'which one?' is ignored in favour of this presupposition. Are the pair of sea urchin cells produced by the first cell division in embryogenesis the two halves of one urchin or two individuals? This question cannot be answered a priori.

The third alternative Ruyer identifies is vitalism, the view that to explain the dynamic character of material reality, one must invoke a superadded 'vital principle'. For Ruyer, vitalism is an obvious recourse in the face of the equipotential characteristics of embryogenesis. As he writes,

> The temptation posed by vitalism or animism is considerably better justified than mechanists commonly think. It does not simply rest on the fanciful belief in a kind of vital breath added to the visible matter of the living being, introduced like the breath of Yahweh into clay formed into the shape of a man. It rests, rather, on the intuition that organic forms are not of the same kind as, or extrapolations of, physico-chemical forms, and that their respective modes of complexity are completely different. (1958: 50)

Vitalism's fatal flaw, though – and one that it shares with the familiar contemporary thematic of 'emergence' – is to accede in advance to the truth of mechanist materialism. It not only agrees that there is a 'basement level' material world that operates according to the laws of macro cause and effect, to which something must be added in order for the former to be explained. It also conceives of this vital surplus *in mechanist terms*. Vitalism in this sense is therefore 'false, not because it attributes the character of a force to the mind or the vital direction, but because it endows it with the character of a macroscopic force that could act directly on phenomena at our human scale and directly realize an intention' (2016: 6).

The solution will be to reject the distinction between the inanimate and the animate in the opposite direction, and at the same time to undermine the vitalist's 'intuition' of a difference in kind between organic and physico-chemical forms – by accounting for all genuine beings, from atoms and molecules up, as conscious, living individualities. Or, as Deleuze puts it, arguably invoking Ruyer as he inverts Leibniz: 'Every spatio-temporal dynamism is accompanied by the emergence of an elementary consciousness which itself traces directions, doubles movements and migrations, and is born on the threshold of the condensed singularities of the body or object whose consciousness it is' (1994: 220).

## *A generalised embryogenetic ontology*

Ruyer's own neo-finalist philosophy holds that the mechanism of macrophysical accounts of reality is most profoundly at error because it rules out any recourse to a formal constituent in formative processes. For Ruyer, such an element *must* be in play in order to account for even the most basic maintenance of individuality – how a benzene molecule holds together, or an amoebic colony, or a joint work project, whether of ants, bees or human beings – let alone the formation of a new individual, the genesis of a living form. He will give this factor – what *informs* dynamic ontogenetic processes – a variety of names, including form, domanial unity, and absolute surface, but the key name here is *thematic development*. Every individual being expresses and is formed in accordance with such themes. For Ruyer, these virtual (his own word) themes are not fixed structures directly applied to material reality like a mould to an inanimate matter,[13] but, precisely like a musical theme or refrain, possess an ideal elasticity, are capable of being played out in different keys, registers and timbres, and at different, even varying, speeds. The analogy can only carry us so far, but it is illuminating. Leibniz gives us the fixed pianola of being; Ruyer, the always singular performance of a song long sung.

What does this mean for embryogenesis? On the one hand, it allows Ruyer to account for the dramatic dynamisms involved without giving up the role of an ideal or guide to development, and without giving in to brute macrophysical causation. As he puts it at one point in *La genèse des formes vivantes*, 'Segmentation, bilateral, radial or bifurcated symmetry is subordinated to a morphogenetic theme whose action is not explicable in mechanical terms' (1958: 87). So the various positions through which Dalcq oscillates, and the field-gradient-threshold around which the other modern embryological thinkers Deleuze cites circulated, are displaced, becoming particular *results* rather than the *source* of action. It is not, therefore, that Ruyer disagrees with Dalcq's explicit thesis about the existence of a dynamic source of biological organisation, but only that the latter's account of this source is impoverished; it is this poverty that leads Dalcq to the oscillation Ruyer diagnoses between a physiochemical mechanism, a poorly conceived finalism and an obscurely conceived vitalism. On the other hand, and because for Ruyer these themes pertain not just to the embryonic moment in organic development, but to all beings from atoms in their play to human beings in the throes of a moral dilemma, embryogenesis is itself another name for ontogenesis *tout court*.

## *Deleuze with Ruyer: intensive individuation*

My third and final point here concerns the closeness of Ruyer's vision of embryogenesis with Deleuze's ontology of intensive individuation. Like Ruyer, the Deleuze of *Difference and Repetition* thoroughly rejects the materialist position. This is true not only of its mechanist aspect, however finessed by thermodynamics, but in its vitalist aspect too. There is a non-material, non-actual facet of reality that constitutes the 'sufficient reason' for the advent of the new.[14] This is the virtual, which Deleuze not only calls Idea but also *theme*: 'an Idea is a "complex theme", an internal multiplicity – in other words, a system of multiple, non-localisable connections between differential elements which is incarnated in real relations and actual terms' (1994: 183). That 'complex theme' appears here in quotation marks indicates, I would argue, an invocation of Ruyer. When the latter appears explicitly, let's recall, Deleuze writes that cellular migration (i.e., a change in or development of an organism) involves 'a structural "theme" to be actualised' (216). In terms of embryogenesis, we can say that the advent of new beings – mutations in the realm of extended and qualified reality – does not begin at the level of matter qua extended, discrete particles. Its *ratio* is to be found at the ideal level of the virtual qua thematic.

Again like Ruyer, Deleuze rejects the alternative presented by preformationism. The Idea is not an eternally unchanging ideal, but this differential structure. In terms of embryogenesis, we can say that the advent of new beings cannot be accounted for by reference to fundamental kinds (categories or species). The ideal component in embryogenesis is a multiplicity, a differential field that problematises the existing distribution of extension and quality, not an ideal ipseity.

Finally, and perhaps most controversially, it seems to me that Deleuze, again like Ruyer, does not succumb to the 'vitalist temptation' (1958: 50), if by this we mean to invoke a vitality that either *emerges from* matter or is *added to it*. The explicit terms of Deleuze's self-ascribed vitalism are worth keeping in mind here, especially the most well-known remark, one often only cited in part: 'Everything I've written is vitalist, at least I hope it is, and *amounts to a theory of signs and events*' (1995: 143, translation modified and emphasis added). Especially given the ideal character of events in Deleuze's work, we are here a long way from the tradition of *Lebensphilosophie*, and arguably at some distance from the runnels of new materialism's vital matter.[15] For Deleuze, in a word, what explains the advent of the new is not the restless character of materiality – its intensive facet – but the expression of an ideal event or theme *through* intensity. Lose the theme, and everything else comes to naught. Looking over these three critical points, we can see that they involve the three registers of the ontology of *Difference and Repetition*: the actual (materialism), the virtual (preformationism) and the intensive (vitalism).[16] In all three respects, we see a clear affinity with Ruyer's position.

## Darwin Between Deleuze and Ruyer, Deleuze Between Darwin and Ruyer

I would like to conclude by returning to Darwin. In *Difference and Repetition*, as we have seen, Darwin appears in the guise of a thinker of individual difference – an ill-fitting costume. The actual project of the book, on the other hand, is much closer to Ruyer's thought as I have outlined it here. Every genuine being can certainly be described as one member of a crowd, a discrete piece in an environmentally situated mass, but this tells us nothing about the ontogenesis of that being – which is to say, it tells us almost nothing about that being at all, neither its origins, nor its ongoing capacity to become, to be other than what it is in the present. In other words, the embryo is not just an example for Ruyer or for Deleuze because everything arises through embryogenesis, expresses a virtual developmental theme, and retains an intensive equipotentiality.

Now, as Ruyer himself insists, there can be no question of jettisoning Darwin's work in favour of another non-evolutionary account.

> By critiquing the theory of natural selection, we are thus not replacing one theory with another, an anti-finalist unilateral theory with a finalist unilateral theory. Instead, it is a question of introducing at least one factor that has a finalist direction, which operates according to very diverse modes and which has to be combined with the other factors highlighted by the neo-Darwinians or the anti-finalist biologists. (2016: 168)

The factor that must be introduced can be neither a fixed form (*de facto* finalism, preformationism), nor a superadded vital principle – the first possessing 'too much' fixity and the latter 'not enough' formative character. The additional factor is the vector of improvisation and change: Ruyer's form, and Deleuze's virtually structured intensive individual.

But the analysis advanced here should not be taken to conclude that Deleuze and Ruyer's respective accounts enjoy a completely harmonious relationship. While this is not the place to examine this point in any detail, I will hazard one final remark.[17] Ruyer's neo-finalism not only insists on the individual as the sole ground for any ontology, he correlatively rejects any ontological significance of environmental factors, rejects the epigenetic in the same gesture as the genetic (genetics being for him 'just a new avatar of preformationism' [2016: 182]). For Deleuze, though, the Leibnizian principle of closure that Ruyer so clearly embraces also constitutes a forgetting, this time of the openness of the intensive individual and its implication with the entire field of infra-being. This is to say that Deleuze *embraces* both the genetic and the epigenetic in the same gesture, while, in the wake of Ruyer, reformulating them in dynamic, ontogenetic terms. Thus he writes:

> A living being is not only defined genetically, by the dynamisms which determine its internal milieu, but also ecologically, by the external movements which preside over its distribution within an extensity. A kinetics of population adjoins, without resembling, the kinetics of the egg; a geographic process of isolation may be no less formative of species than internal genetic variations, and sometimes precedes the latter. (1994: 216–17)[18]

And here we find ourselves back in the company of Darwin: not the counter-sensical Darwin that Deleuze explicitly presents, but Darwin the naturalist, Darwin the thinker of the milieu and the pack. From this point of view, Ruyer and Darwin are no longer opposed, but continue each other's insights in their respective registers. If it is a matter of adding something 'vertical' to Darwin's account of natural selection, as Ruyer thinks, it is equally important – if we want to understand why Deleuze makes use of biological thought and how he does so – that we supplement Ruyer's contemporary Leibnizian thought with something 'horizontal', collective and environmental that has its profound roots in the thought of Darwin: an understanding of 'effective multiplicities' (Sauvagnargues 2014: 416), or what Darwin will call the 'swarm' (1859: 202).

## Notes

1. I leave aside here the following remark of Darwin's: 'Nor shall I discuss the various definitions which have been given of the term "species." No one definition has satisfied all naturalists; yet every naturalist knows vaguely what he means when he speaks of a species' (1859: 38), and the proximity of this formulation to the 'Everybody knows that' attacked by Deleuze in the 'Image of Thought' chapter of *Difference and Repetition*.
2. Even leaving aside the fact that Darwin very often defines the individual in terms of the species, there is good reason to think that he did not view the species level as completely nominal; that, as Gordon Chancellor puts it, 'Darwin recognised supraorganismal selection and struggled to understand its implications'; and that he ultimately arrives at some form of multi-level selectionism that includes the species as one of its real, if not unique, levels (Chancellor 2015: 138). Chancellor's paper is a good overview of the reasons to assert a 'reduction to the organism, plus . . .' reading of Darwin's position; see too, of course, Gould 2002: 64, and fundamentally *passim*. This is of marginal interest in the current context, however, so I leave it aside. I also leave aside the brief critical construction of nominalism by Deleuze himself in the opening pages of *Difference and Repetition*, according to which it appears as a kind of negative realism (in the sense that one speaks of 'negative theology').
3. Broadly speaking, these passages of *Difference and Repetition* deal with the same material as the 'Geology of Morals' plateau in *A Thousand Plateaus*; the facet of the argument that involves the 'trinity' of Cuvier, Geoffroy and Darwin (and Baer in passing) is also discussed by Deleuze in one of his lectures on Foucault and power in 1986: <http://www2.univ-paris8.fr/deleuze/article.php3?id_article=464> (last accessed 29 May 2017).
4. Deleuze, 1994: 216: 'The entire world is an egg'; 'The world is an egg, but the egg itself is a theatre' (251); 'The world is an egg' (251).
5. Regarding this other trajectory that links together Cuvier, Saint-Hilaire, Perrier and Darwin, see Somers-Hall 2013: 212–33; see too the very clear and useful page of Ansell-Pearson 1999: 159–60.
6. Deleuze's conflation of von Baer and Vialleton's respective positions can be found at 1994: 249.
7. The concept of field was first developed by Gurswitch and Weiss, and gradient by Child.

8. Such an analysis would have to begin with Henning Schmigden's observation that many of Deleuze's biological references, including Dalcq, appear together in the annotated bibliography of Georges Canguilhem's *La connaissance de la vie* (Schmidgen 2015: 123–49).
9. In this, Dalcq's theory is of a piece with the trajectory of the notion of field in developmental biology before him, for instance in Huxley and de Beer: 'The determination and localisation of organ-rudiments is revealed sooner or later by the presence of chemodifferentiated material or morphogenetic substances in certain places which constitute what may be called *fields*, or areas of differentiation of organ' (cited at Waddington 1966: 107). Waddington's concern with the vagueness of this idea is easy to sympathise with; Deleuze addresses the same problem of vagueness in his definition of an intensive field, right after his discussion of Dalcq (1994: 251). His response turns around the way in which intensive individuals express the differential structure of the virtual, which can be very precisely theorised (as Deleuze shows) by way of the differential calculus.
10. I owe this way of framing the point to Ronald Bogue: 'If "the world is an egg," as Deleuze asserts in *Difference and Repetition*, it is through an examination of the thought of Ruyer and its appropriation by Deleuze that one may most easily grasp the full implications of this assertion' (2009: 300).
11. Chapters 1 and 3 of *La genèse des formes vivantes* deal with each of these points in turn. Here, Ruyer is clearly (and explicitly) in the lineage of the embryologist Hans Spemann, who advances a direct attack on the field-gradient analysis begun by Thomas Hunt Morgan. For Spemann's critique of the field-gradient approach, see Spemann 1938; cf. Thieffry 2001: 151. This mode of analysis, as we have seen, was popularised by Child and given a broad-ranging scope in Dalcq's work. These concepts do not, as Deleuze seems to indicate at 1994: 251, originate in Dalcq's work but in Thomas Hunt Morgan (see Thieffry 2001: 152).
12. We find here a (less witty) biological analogue to the infamous remark that 'the working mathematician is a Platonist on weekdays and a formalist on Sundays' – in the lab, the embryologist is a *de facto* finalist, an implicit preformationist; in the classroom, a chemico-physical mechanist. Driesch's shock at the equipotentiality of the sea urchin embryo is the shock of having this implicit belief overturned (see Ruyer 1958: 88–9).
13. This Aristotelian hylomorphism is directly taken up and criticised by Ruyer under the heading 'Pseudo-formation' (1958: 12ff.).
14. On this conjunction, see the classic article by Daniel W. Smith (2008).
15. I agree with Claire Colebrook's assessment on this point (see Haynes 2014). But see too Woodward 2015: esp. 26–8.
16. I use the term *register* here in a very general sense; it seems to me in fact that the category of the actual includes both intensive, dynamic processes and the extended and qualified result that is often described by Deleuze's commentators under the heading of the actual. On this, see Roffe 2012, especially the second half of chapter 7.

17. Critical reflections of this kind are just now, at the time of writing, beginning to appear. The most important instance at present is Anne Sauvagnargues' reflection on *Neofinalism* (2014: 402–16). I criticise Ruyer from a complementary angle (2017). There is also arguably a question around the adequacy of Ruyer's distinction of what Deleuze calls the virtual and the intensive which will be pursued in further publications.
18. It is notable that Deleuze appends to this passage a note invoking neither Ruyer nor Darwin, but Geoffroy Saint-Hilaire, and this account resembles less Ruyer's neo-finalism than it does Gilbert Simondon's philosophy of transduction. The *braiding* of texts thus continues.

## References


Ansell-Pearson, K. (1999), *Germinal Life*, New York: Routledge.

Bogue, R. (2009), 'Raymond Ruyer', *Deleuze's Philosophical Lineage*, ed. G. Jones and J. Roffe, Edinburgh: Edinburgh University Press, pp. 300–20.

Chancellor, G. (2015), 'Levels of Selection in Darwin's *Origin of Species*', *History and Philosophy of the Life Sciences*, 37: 2.

Churchill, F. B. (1991), 'The Rise of Classical Descriptive Embryology', *Developmental Biology*, vol. 7, ed. S. F. Gilbert, New York: Plenum Press.

Darwin, C. (1859), *On the Origin of Species by Means of Natural Selection, or the Preservation of Favoured Races in the Struggle for Life*, London: John Murray.

Deleuze, G. (1988), *Le pli: Leibniz et le baroque*, Paris: Minuit.

Deleuze, G. (1994), *Difference and Repetition*, trans. P. Patton, New York: Columbia University Press.

Deleuze, G. (1995), *Negotiations*, trans. M. Joughin, New York: Columbia University Press.

Gould, S. J. (2002), *The Structure of Evolutionary Theory*, Cambridge, MA: Harvard University Press.

Haynes, P. (2014), 'Creative Becoming and the Patiency of Matter: Feminism, New Materialism and Theology', *Angelaki* 19: 1, pp. 129–50.

Hyman, L. H. (1957), *Charles Manning Child, 1869–1954*, Washington: National Academy of Sciences.

Roffe, J. (2012), *Badiou's Deleuze*, New York: Routledge.

Roffe, J. (2017), 'Form IV: From Ruyer's Psychobiology to Deleuze and Guattari's Socius', *Deleuze Studies* 11: 4, pp. 580–99.

Ruyer, R. (1958), *La genèse des formes vivantes*, Paris: Flammarion.

Ruyer, R. (2013), *L'embryogenèse du monde et le Dieu silencieux*, Paris: Klincksieck.

Ruyer, R. (2016), *Neofinalism*, trans. A. Edlebi, Minneapolis: University of Minnesota Press.

Sauvagnargues, A. (2014), 'L'averse de sable, l'atome et l'embryon', *Critique* 804, pp. 402–16.

Schmidgen, H. (2015), 'Cerebral Drawings Between Art and Science: On Gilles Deleuze's Philosophy of Concepts', *Theory, Culture & Society*, 32: 7–8, pp. 123–49.

Smith, D. W. (2008), 'The Conditions of the New', *Deleuze Studies* 1: 1, pp. 1–21.
Somers-Hall, H. (2013), *Hegel, Deleuze, and the Critique of Representation*, Albany: State University of New York Press.
Spemann, H. (1938), *Embryonic Development and Induction*, New Haven: Yale University Press.
Thieffry, D. (2001), 'Rationalizing Early Embryogenesis in the 1930s: Albert Dalcq on Gradients and Fields', *Journal of the History of Biology*, 34: 1.
Vucinich, A. (1988), *Darwin in Russian Thought*, Berkeley: University of California Press.
Waddington, C. H. (1966), 'Fields and Gradients', *Major Problems in Developmental Biology*, ed. M. Locke, New York: Academic Press.
Weiss, P. (1973), *The Science of Life*, Mount Kisco: Futura Publishing.
Woodward, A. (2015), 'Nonhuman Life', *Deleuze and the Non/Human*, ed. J. Roffe and H. Stark, Basingstoke: Palgrave, pp. 25–41.

Chapter 3

# Framing Sexual Selection: Elizabeth Grosz's Work on Deleuze, Darwin and Feminism

*Erin Hortle and Hannah Stark*

Elizabeth Grosz is arguably one of the most important feminist philosophers of the early twenty-first century.[1] Her exuberant work has offered a deep and sustained consideration of space, time and bodies in relation to sexuality and sexual difference. In Grosz's recent publications she has taken up the work of Charles Darwin which contains a set of philosophical tools for revitalising the topics with which she had previously been concerned. Two thinkers are foundational to Grosz's Darwinian project: Gilles Deleuze and Luce Irigaray. These three very different thinkers – Darwin, Deleuze, Irigaray – sit in Grosz's work like the sides of a triangle, each positioned so that they hold the others in place. As a range of commentators have suggested, Grosz uses Darwin to negotiate the divide between Deleuze's ontology of pure difference and Irigaray's ethics of sexual difference (Pulkkinen 2017; Parisi 2010; Weinstein 2010). For Grosz, as for Irigaray, sexual difference is a foundational difference both philosophically and materially. Their projects are commensurate in that they both assert that thinking sexual difference is politically important for countering a neutral (masculine) model of the body with the specificity of sexually different morphologies. However, this is not to say that their theorisations of the generative capacities of embodied sexual difference are entirely the same.[2] In *Becoming Undone*, Grosz writes that life 'is the elaboration of at least two lines of development, two morphologies, two types of body: a divergent development that brings with it endless variation and endless difference' (2011: 3). Bringing Deleuze and Irigaray together in a feminist evolutionary project is not a simple matter, for as Grosz herself acknowledges, Deleuze and Irigaray 'generate a tension, their concepts do not fit together well' (2008: x). For Deleuze, difference is the prior ontological condition which sits behind the appearance of any organism, identity or category. This is the pure difference that Deleuze describes as 'a swarm of differences, a pluralism of free, wild or untamed differences' (2001: 50), the difference that can never be reduced to 'identity, opposition, analogy and resemblance' (29). This is difference in and of itself, and which exists as pure positivity. As Pulkkinen writes, 'if sexual difference is a difference that is irreducible, precisely on an ontological level, then the Deleuzian ontology of constant flow of differences and identities into others seems to meet a limit, a point where the foundational, irreducible difference of masculinity and femininity does not flow' (2017: 283). In emphasising the irreducible centrality of sexual difference to Irigaray's work, Pulkkinen points to its incommensurability with the flux of difference on which Deleuze's ontology is founded. With this in mind, why would Grosz turn to Deleuze as one of her major influences for a feminist evolutionary project which situates sexual difference as the primary difference for ontology and for philosophy?

Mobilising Deleuze in any evolutionary project is fraught. Deleuze and Darwin share a common endeavour, Nathan Eckstrand writes, 'to reject a typology that classifies entities into universal, absolute categories and propose a system by which new beings and new differences can come into existence' (2014: 416). Deleuze, like Darwin, is interested in how change occurs over time, as is evident, most notably, with his concept of becoming. However, Deleuze is very careful to distinguish becoming from evolution. With Guattari he writes 'becoming is not an evolution, at least not an evolution by descent and filiation' (2004: 263). Deleuze and Guattari characterise evolution in Darwinian terms, as a process that works through 'genealogy, kinship, descent and filiation' (258), that is, a linear movement of hereditary material through time. Alternatively, becoming, produced when disparate things like the wasp and the orchid come together to make something new, is about alliance. The nuptials of the wasp and the orchid involve entities which are radically different; while this trans-species and trans-kingdom alliance will produce something new it will not be a wasp-orchid child.

In *Difference and Repetition*, Deleuze applauds Darwin for his commitment to individual difference which, in *The Origin of Species*, creates the conditions necessary for natural selection to occur in a population. Natural selection shows both how differences come together to congeal into a particular shared trait, and also how differences spread apart and enable things to differentiate from each other (2001: 248). Deleuze acknowledges in this passage that the connection between natural selection and sexual reproduction facilitates the proliferation of individual differences (249). However, for Deleuze, difference occurs not within a species but prior to the existence of species as a category. From pure difference, individuation produces individuals which then get sorted into categories such as species. Deleuze writes:

> It is not the individual which is an illusion in relation to the genius of the species, but the species which is an illusion – inevitable and well-founded, it is true – in relation to the play of the individual and individuation. The question is not whether in fact the individual can be separated from its species and its parts. It cannot. However, does not this very 'inseparability', along with the speed of appearance of the species and its parts, testify to the primacy in principle of individuation over differenciation? It is the individual which is above the species, and precedes the species in principle. (250)

The major distinction between Deleuze and Darwin, then, concerns the location of the generation of difference. For Darwin difference emerges within a species and thus enables the selection of those traits that are more fit to a particular environment. This is not to say that 'species' is in any way a simple or unproblematic category for Darwin, only that in his work it is a location where difference manifests itself and is important to evolution. For Deleuze, alternatively, difference was always prior to species or to the individual. As Eckstrand suggests, for Deleuze, 'species are an illusion patterned onto the appearance of individuals via individuation' (2014: 420), and this is why we must, on Deleuze's account, 'deal first with the conditions for the appearance of the singular individual and second with the individual's categorisation' (419).

The tension evident here, between the individual and the species, can be seen in the distinction, in Deleuze's work, between established structures of coherence and the destabilising capacities of genetic difference and molecular becomings. For Deleuze, structure does not precede genesis but emerges as a way of organising a world in which difference is tantamount. Deleuze's pure difference ensures that we cannot give an ontological or essential status to particular structures or identities. For Deleuze, the individual, as a unit of pure difference, is prior to any category into which it can be organised. As Arun Saldanha reminds us in relation to race, a human body is 'unique in its morphology, coloration, affects and memories, individuating itself by feeding on all sorts of inhuman elements (DNA, books, money, ships, horses), which it mostly shares with other bodies. Only secondarily does a body become lumped together with other bodies, to give rise to populations circumscribed on the basis of perceived similitude and proximity' (2013: 10). However, this is not to say that identity does not exist for Deleuze or that groups based on shared characteristics are invalid. Rather, individuation produces individuals which can *then* be grouped into species; individuation comes first.

In the same way that for Deleuze species emerge from a prior field of pure difference, sexual difference also emerges out of a more primordial difference. Sexual difference is not, then, a primary difference for Deleuze as it is for Irigaray and Grosz. In *Anti-Oedipus* he and Guattari write of biology as an 'intensive germinal flow' (2005: 162) which gets organised into the categories of male and female. Offering an historicised account, they focus on how sexual difference came to be an organising principle for the social arrangement of bodies. They illustrate this though the distinction between three modes of social organisation or regimes: primitive and territorial, despotic, and capitalist. In what Deleuze and Guattari describe as 'primitive' societies there are first infinite flows of 'women and children, flows of herds and seed, sperm flows, flows of shit, menstrual flows' (142). Bodies are initially organised genetically into tribes and from here they are distributed in relation to social codes that endeavour to organise or territorialise these excessive germinal flows in order to produce the social field. When we arrive at capitalism the organisation of bodies in relation to these social codes no longer makes sense. In this system, economic codes are not fixed and neither are patterns of exchange. Gendered roles within the family in a capitalist society emerge through a specific relation to the Oedipus Complex, which organises desire into a set pattern. Shifts in the organisation of bodies between the primitive, despotic and capitalist regimes tell us that their current arrangement is socio-historically specific and therefore subject to potential reconfigurations. Gendered roles within society are a manifestation of particular social and economic structures and cannot express anything innate or universal. Moreover, sexuality and desire are fundamentally excessive and complex. The place of sex, gender and sexuality within *Anti-Oedipus* indicates that for Deleuze and Guattari bodies, which are infinitely different, are merely territorialised as 'man' or 'woman' by sexual difference. In other words, 'sexual difference' is for them the name of a territorialisation, rather than a deterritorialised flow. In *A Thousand Plateaus* Deleuze and Guattari counter what they describe as a 'great binary aggregate' of the two-sex system with 'a thousand tiny sexes' (2004: 235) which are molecular and mobile.[3] Despite her commitment to an Irigarayan model of sexual difference, it is this point – that sexes are molecular and mobile – that sheds most light on how Grosz draws upon Deleuzian thought to elaborate her feminist ontology of sexual difference.

Grosz's ontology of sexual difference is an evolutionary project that starts at the beginning, or rather, 'with some mythical sense of "the beginning"' (2008: 5). This 'beginning', Grosz admits, is a simplified Deleuzian image of the beginning (see 2011: 77). It starts with 'chaos, the whirling, unpredictable movement of forces, vibratory oscillations that constitute the universe' (2008: 5). Chaos for Grosz is not absolute disorder but rather 'a plethora of orders, forms, wills . . . both matter and its conditions for being otherwise, both the actual and the virtual indistinguishable' (5). The version of chaos that Grosz offers can be read as roughly analogous to Deleuze's vision of the pure difference that subsists beneath the emergence of specific things. Randomly occurring in this chaos is a molecular happening: life emerges. In its emergence, this life extracts from chaos the elements, substances or processes it requires to differentiate itself, which it does by simultaneously enframing itself and 'cast[ing] into shadow the profusion of forces that engulf and surround it' (6). In other words, life emerges as something distinct through a process of individuation by territorialisation. Life is one of the plethora of orders that exists in chaos, but crucially it carries with it the ability 'to *act*, to enfold matter into itself, [and so] to transform matter and life in unpredictable ways' (6, emphasis added). This agency is life's method of replicating or reproducing itself.

Over time, for some organisms, asexual reproduction split into a more elaborate form of sexual reproduction, and hermaphroditism and androgyny into sexual bifurcation. For Grosz, sexual difference is what guarantees the proliferation of concrete differences in the world. In defence of the place of sexual difference in Irigaray's project, and of her use of Irigaray in her own, Grosz writes:

> [w]ithout sexual difference, there could be no life as we know it, no living bodies, no terrestrial movement, no differentiation of species, no differentiation of humans from each other into races and classes – only sameness, monosexuality, hermaphroditism, the endless structured (bacterial or microbial) reproduction of the same. (2011: 101)

Grosz's reading of the significance of the shift to sexual bifurcation draws heavily on the writings of Darwin who, in speculating upon the origin of certain vestigial or rudimentary organs of one sex being found in the other (such as the presence of nipples in males), suggests that 'some extremely remote progenitor of the whole vertebrate kingdom appears to have been hermaphrodite or androgynous' (2010: 207). In order to excavate how and why this shift from hermaphroditism and androgyny into sexual bifurcation occurred, and what it effected, Darwin spent eight years examining barnacles. Grosz uses his work on barnacles as a case study for her essay 'The Nature of Sexual Difference', in which she seeks to bring Irigaray's theory of sexual difference into conversation with Darwin's theory of sexual selection. She speculates that '[i]t is perhaps his fascination with th[e] question of sexual difference . . . that directed him to the laborious investigation . . . of the evolutionary history of the barnacle' (2012: 85).

Darwin discovered that many barnacle fossils have a hermaphroditic anatomy, and that while contemporary barnacles are mostly sexually bifurcated, some quite rare forms also appear to be hermaphrodites, although the male organs of these hermaphrodites are 'in a process of atrophy and are "microscopically small"' (86). Moreover, these overwhelmingly female hermaphrodites have small parasites attached to them. Grosz recounts Darwin's discovery that the parasites are actually dwarf males in the process of emerging as a separate sex but, as yet, she writes, they 'are still largely primitive modes of insemination, little more than sacks of sperm' (86). Thus, in one single species, 'there are hermaphrodites, which are largely female, there are atrophying male organs, and tiny self-contained males' (86). Darwin explained this phenomenon as resulting from the morphological limits of the barnacle's life. Grosz summarises:

> Sessile or immovable animals, such as barnacles . . . grow their shell directly on to a fixed place, and after this attachment they are not able to move. Thus barnacles can only exchange sperm with their nearest neighbours. If their nearest neighbour is one of two sexes, then there is only a 50 per cent chance that a male and female will be in proximity with each other. With two hermaphroditic individuals, the possibilities for cross-fertilisation are greatly enhanced . . . And the emergence of complemental or parasitic males . . . which attach themselves to the female/hermaphrodite . . . is also much more likely to result in cross-fertilisation than two separate and autonomous sexes. (86)

For Grosz, following Darwin, the question this scenario both asks and answers is: 'how does this hermaphroditic creature become a sexually differentiated creature?' (86). These types of barnacles, Grosz reasons, represent a transitional stage in sexual bifurcation: 'here hermaphrodites incorporate male organs that are clearly in the process of atrophy, as well as robust female organs, for their function can now be assured fertilization with the emergence of the complemental males' (87). Over time, the hermaphroditic male organs will either disappear completely, or become vestigial, and the process of sexual bifurcation will be complete, or as complete as anything can be when considered through an evolutionary lens.

This split into sexual bifurcation and associated sexual reproduction was an evolutionary shift driven arguably by the forces of natural selection, but in the process, something quite distinct was put in motion. Natural selection, as evidenced in the example of the barnacle, privileged sexual bifurcation over hermaphroditism, because the chances of reproduction were greatly increased. Moreover, and relatedly, natural selection privileged sexual reproduction because an organism or species that sexually reproduces rather than simply replicating itself (as some bacteria and viruses do), allows for greater gene survival through the proliferation and variability in recombinant DNA (Colebrook 2012: 170). Sexual difference, in turn, gave rise to another evolutionary driver in sexual selection, a force that allowed ever more varied forms of life to emerge and perpetually, yet differentially, proliferate, through the sheer force of individuals exercising choice based upon taste when it comes to selecting sexual partners. 'For Darwin', Grosz writes, 'sexual bifurcation, the eruption of more than one (by default, female) sex, is a strategy to maximize the potential for variation, to maximize the forms of living beings, to maximize difference itself' (2012: 87).

Sexual reproduction allows for variation to occur at faster speeds than does asexual reproduction, thus compounding the proliferation of difference in the world in a material sense. In other words, sexual difference causes nature to be unstable, or at least more unstable than it was. Grosz argues that '[n]atural selection selected the strategy of dividing the sexes into (at least) two bodies rather than a single morphological type – not for all the forms of life, especially those invested in evolutionary stable environments, *but for those involved in changing situations*' (87, emphasis added). It is for this precise reason (and in a move informed in no small part by Irigaray's claim that sexual difference is *the* central concept of our age [Irigaray 2005; Grosz 2012: 70]) that Grosz suggests that the development of sexual difference is 'one of the earliest upheavals in the evolution of life on earth and undoubtedly *the most momentous invention that life has brought forth*, the very machinery for guaranteeing the endless generations of morphological and genetic variation, the very mechanism of biological difference itself' (2008: 6, emphasis added). The weight of Grosz's words here is telling, and her claim signifies the juncture at which her theorisation of difference splits with Deleuze and takes up, more closely, with Darwin. While Deleuze allows that the energies of sexual reproduction facilitate the proliferation of individual differences (2001: 249), for him, pure difference remains the primary method of individuation. For Grosz, pure difference is always virtually present in the chaos that surrounds, comprises and is comprised of individuals and species, but it is sexual selection (since the emergence of sexual difference) that actualises the proliferation of those differences on earth. Sexual selection is an evolutionary process that does not appear to interest Deleuze. This is noteworthy, because sexual selection 'deflect[s] natural selection through its extravagant and excessive pleasures, its inventions and intensification of new relations, new forms of attraction and new modes of artfulness', and thus, Grosz argues, 'it is responsible for the intensification of beauty over generations and for the proliferation of colors, sounds, forms that are pleasing to members of one's own species if not others' (2012: 87). Because sexual selection is geared toward tastes within species (and even, Grosz suggests, tastes across species), it has a significant effect on the material world.

This is a point we will return to shortly, but first we want to address more closely the distinction between natural and sexual selection which is so central to Grosz's theorisation. While natural selection gave rise to sexual selection, sexual selection is a force fundamentally irreducible to natural selection. In this way, the natural selection of sexually dimorphous barnacles over those that were hermaphroditic causes an evolutionary shift. However, when these barnacles reproduce sexually and the speed at which the manifestation of species diversity increases, those barnacles whose differences are best suited to the environment will be selected to reproduce their genes sexually. Natural selection, Grosz synopsises, 'is always ultimately directed by the struggle for existence, the struggle to survive: it is oriented to the factors regulating life and death' (81). Sexual selection, on the other hand, is driven by the struggle to allure and acquire sexual partners (81); thus, as Darwin argues, sexual selection 'depends, not on a struggle for existence, but on a struggle between the males for possession of the females' (1998: 117).[4] While sexual selection is, in some regards, 'subordinate' to natural selection (Grosz 2012: 81), and as Darwin acknowledges, it is certainly 'less rigorous than natural selection' (for 'the result is not death to the unsuccessful competitor, but few or no offspring' [1998: 117]), as Grosz, following Darwin, takes pains to stress, sexual selection is a force distinct from natural selection (2012: 81).

This distinction is most apparent when one considers the ways in which sexual selection not only complicates but also undermines the processes of natural selection. As behaviours and bodies co-opted by the purview of sexual selection have become more and more elaborate, sexual selection has developed the potential to imperil life. The more ornate bodies and behaviours are, the more obvious they are, and the more vulnerable they are to attracting not only potential sexual partners, but also predators (81). The example Grosz draws upon to illustrate this point is the peacock. Why would natural selection produce such magnificent plumage? For the more beautiful the peacock's plumage, the more visible he is, not only to peahens and other peacocks, but to potential predators (2008: 30). 'Sexual selection', Grosz argues, 'imperils as much as it allures . . . it generates risk to the same extent that it produces difference' (30). Although 'arguably the greatest invention of natural selection' (2012: 87), sexual selection does not necessarily observe its rules, and, Grosz argues, in affirming the distinctness of sexual selection, 'Darwin introduces an excessiveness into the development and transformation of species' (2008: 33) that natural selection alone cannot account for.

The importance of this irreducibility for Grosz's feminist project cannot be stressed enough; and it is for this reason that, despite Darwin's complex relation to feminism, she persists in using his theories to inform and serve her ontology of sexual difference. As Parisi writes, 'Grosz specifically suggests that theories of evolution can help feminism to re-address biological difference not as a social, cultural, or political limit, but as providing a compelling challenge to essentialist views of sex' (2010: 151). The notion of sexual selection as an irreducible force underpins Grosz's radical rearticulation of the biological and natural, which, she suggests, were co-opted by Darwin's contemporaries and followers in their misreading of sexual selection. According to Grosz, the continued affirmation of the difference between sexual and natural selection is the most major distinction between Darwin's theories and those of contemporary sociobiologists, for whom sexual selection is 'a subtle form of natural selection, and attractiveness . . . an index for fitness' (2012: 81). Grosz argues that over the ensuing centuries evolutionary psychologists took up an approach to sexual difference which worked to reduce 'maleness and femaleness to their reproductive capacities and activities alone' (79) and as such contributed to a naturalisation of hetero-patriarchy. However, this broad reading of Darwinian evolution is, Grosz argues, a gross misreading (79), and, we would add, it is also a gross misreading and misappropriation of the differences and excesses observable in the natural world.[5]

Absurdly, such misreadings of sexual selection (and relatedly, of Grosz's own work) as affirming hetero-patriarchy actually espouse an inverted understanding of sexual selection by privileging reduction over excess. Moreover, they fail to attend to the fact that sexual activities and practices are not only and not always geared toward reproduction, but toward something more. It is through this attention to excess that Grosz finds in sexual selection, and which she elaborates so fully, that she avoids a naturalisation of heterosexuality. Sexual selection operates in excess of survival (2011: 118). While sex is a mechanism that enables reproduction, it is not limited to this function, and while selecting sexual partners might be about transmitting genetic material, it cannot be limited to this process. For Grosz, sex is also about the 'irrational, non-functional, nonadaptive operations of sexual selection' (128) such as attraction, pleasure and excitation. Grosz asserts that homosexuality in human and non-human species is part of the general excess that sexual selection generates (an excess that is also evident in non-reproductive heterosexual intercourse) (130).

This is not to say that the existence of same-sex attraction and practices actually prevents reproduction or evolution at the level of species. Moreover, there are cases in which same-sex attraction is caught up in reproductive processes. One might, for instance, consider the Laysan albatross. Researchers studying the Laysan albatross in Hawaii observed that 31 per cent of pair-bonded albatross were same-sex couples (predominantly female) that 'courted, allopreened and shared parenting responsibilities' (Bailey and Zuk 2009: 443). Like many birds, the Laysan albatross tend to favour monogamy, and although same-sex coupled birds do not enjoy the reproductive success of their heterosexual counterparts, same-sex coupled birds are more likely to successfully rear a chick than unpaired females (443). As Nathan W. Bailey and Marlene Zuk argue, because the sex ratio of this colony of birds was 'heavily female biased', non-heterosexual practices positively 'impact the evolutionary dynamics of the population' (444). Same-sex attraction, rather than inhibiting reproduction, facilitates its success. This function of same-sex practices is produced, in part, by natural selection, but such a rationalisation of same-sex practices does not take into account that which is produced by the excesses of sexual selection. These excesses include things like colourful plumage on male birds or the musicality of mating calls; they are what make life artistic, creative and pleasurable.

For Grosz, erotic attraction also acts to increase or intensify difference in that it concerns the specificity of one particular partner over others.[6] In fact, while Grosz acknowledges the way in which sexuality 'carries the burden of biological, cultural, and individual constriction', it is an 'open intervention' which is about neither free-will nor determinism but rather creativity and the expression of 'one's being' (2011: 73). As such, sexual selection, Grosz warns, 'is not itself to be conflated with sexual reproduction'; for while the most measurable form of sexual success is the generation of offspring, the aim of sexual selection is not offspring but sexual relations (2008: 29). From such a vantage, Grosz argues,

> sexuality itself is not so much to be explained in terms of its end goals (which in sociobiological terms are assumed to be the [competitive] reproduction of maximum numbers of [surviving] offspring, where sexual selection is ultimately reduced to natural selection) as in terms of forces, its effects (which can less contentiously be understood as pleasure in interdeterminable forms), which are forms of bodily intensification. (33)[7]

Take, for example, a creature that fascinated Deleuze and Guattari, the *Scenopetes dentirostris*, or Tooth-billed Bowerbird. In *What is Philosophy?*, they write:

> Every morning the *Scenopetes dentirostris*, a bird of the Australian rain forests, cuts leaves, makes them fall to the ground, and turns them over so that the paler internal side contrasts with the earth. In this way it constructs a stage for itself like a ready-made; and directly above, on a creeper or branch, while fluffing its feathers beneath its beak to reveal their yellow roots, it sings a complex song made up from its own notes and, at intervals, those of other birds that it imitates; it is a complete artist. This is not a synaesthesia of the flesh but blocs of sensations in the territory – colors, postures, and sounds that sketch out a total work of art. These sonorous blocs are refrains; but they are also refrains of posture and color, and posture and colors are always being introduced into refrains: bowing low, straightening up, dancing in a circle and a line of colors. The whole of the refrain is the being of sensation. Monuments are refrains. (Deleuze and Guattari 1994: 184, quoted in Grosz 2008: 12)

Here, Grosz suggests, the bird creates not only a territory, but a sexualised territory, a 'space that is [its] own in which [it] can enact a sexual seduction, extract sexual satisfaction, and intensify sexual forces' (2008: 12). This territory, this theatre and display, are brought forth by sexual selection. 'Territory', she continues, 'is that which is produced by the elaborate, if apparently useless, activity of construction, attention grabbing, and display that mark most of sexual selection' (12). Sexual selection, conceived as such, is not just about the body but is a force that exceeds the body: it is about the body's position in space; about the body's specific composition of space; and thus about the body entering into an assemblage through its relation with, and composition of, the chaos of materiality. Over generations, sexual selection has created ever more varied bodies; it has also created song, dance and, as exemplified by the *Scenopetes dentirostris*, theatre. This reading of the relation between the body and the space it not only occupies but, through its occupation, composes is, for Grosz, thoroughly informed by Deleuze, for whom 'art is . . . the extension of the architectural imperative to organise the space of the earth' (10). It is for this reason that Grosz argues sexual selection can be conceived of as the evolutionary origin of art; 'art', she suggests, 'is evolutionary in the sense that it coincides with and harnesses evolutionary accomplishments into avenues of expression that no longer have anything to do with survival' (11) or sexual reproduction and filial evolution. Rather, art and its evolutionary driver, sexual selection, propel individuals and species into never-ending processes of becoming.

For Grosz, a nuanced understanding of sexual selection as irreducible to natural selection recognises that a biological conception of sexual difference is not in any way an essentialising conception, because difference is in a continual state of evolution, thus its status must be conceived of as open-ended; within an evolutionary framework, any other conception of sexual difference is a *misconception*. As such, and despite the fact that the association between the two is so well trodden, an ontological approach to sexual difference need not result in a slippage into essentialising doctrine. In this way, the existence of male and female bodies and the arrangement of these bodies into reproductive heterosexual couples that is central to heteronormativity is a cultural arrangement rather than one rooted in Darwinian biology. In Grosz's words, 'nature need not be seen as static or fixed in order to understand that sexual difference characterizes nature and is one of the most striking features of the natural world' (2011: 148), and a striking feature that opens the world to chaos and a multitude of possibilities.

As Tuika Pulkkinen perceptively notes, 'the most interesting characteristic of Grosz's reading of Darwin is that she uses him to highlight the unpredictable character of the future' (2017: 279). To understand both how and why she does this, one must think about the way Deleuzian philosophy underpins her image of evolution. If chaos contains and is comprised of 'both the actual and the virtual indistinguishably' (Grosz 2008: 5), then it must contain the potential for being actualised otherwise; and if life emerges from chaos through an act of territorialisation, it can be subsumed into chaos through a process of deterritorialisation; for a territory always remains in some way open to the chaos from which it draws its force and to which it is infinitely connected (20). Nature is not static or fixed, but contains a multitude of differences and the continual and boundless potential for a multitude more. The present must always be open to the future, because the present not only contains the potential for many futures, but is in the process of making them. Grosz argues that in drawing the distinction between natural and sexual selection, Darwin affirmed that species are not merely 'natural collections or kinds developed to survive and compete, [but] . . . are also the *a posteriori* and ultimately incalculable consequences of sexual taste, appeal and attraction' (33). As Grosz establishes from the outset, life is creative; it has agency – it has the ability to enframe itself, to territorialise itself and then reterritorialise itself anew. Individuals exercise preferential choices and so propel their species into flux: they (re)make sexual difference *through* their participation in sexual selection. What this means is that sexualised bodies are not static or fixed, and nor are their cultural expressions.

Since her work in *Volatile Bodies* in the early 1990s, Grosz's feminist project has called for an intellectual return to the body and also to the natural. Historically these have been difficult sites for feminist politics and philosophy. 'We have devoted much effort to the social, cultural, representational, historical, and national variations in human relations' she writes. 'We now need to develop a more complex and sophisticated understanding of the ways in which natural forces, both living and non-living, frame, enrich and complicate our understanding of the subject, its interior, and what the subject can know' (2011: 86). Sexual selection, as she envisages it, provides Grosz with a framework to rethink a complex set of forces, bodies and relations in ways that problematise the nature/culture distinction, a binary which has been so central to feminist thought. She uses sexual selection as a way to think about the imbrications of art and sexuality (both broadly construed). In theorising how organisms 'exceed the bare requirements of existence' through 'becoming artistic' (2008: 6) she offers a way to think about the excesses of pleasure, desire, beauty, taste, affect, creativity and intensity. These excesses offer her a way out of the biological determinism that a return to theories of the natural world risks.

Grosz does not utilise Deleuze's ideas of species, individuation or sexual difference, nor does she focus in depth on his engagement with Darwin; she is not producing a Deleuzian orthodoxy. Instead, she looks to the rich theorisation of difference in Deleuze's work and uses this to re-engage Darwin in a feminist project. This deviation is significant and is perhaps the reason why she is able to advance her ontology as a feminist politics. Far from causing her to look back to the past, evolutionary theory provides Grosz with a way to orient feminism to the future. Her feminist project is not about epistemology; nor does it take place within a rights-based framework. For Grosz, feminism is a project for ontology, and it needs to be reconciled through an 'immersion in materiality' (2011: 61). Rooted in the materiality of sexual difference, and sexual selection, Grosz theorises a world that is generative and rich with difference. This is a world in which the materiality of sexual difference, in all of its diversity, is not only acknowledged but is also foundational. In this way, the project of feminism is not just providing a social response to patriarchy but requires that we find new ways to create a world in which the masculine does not stand in for the universal human and in which female morphology, experience and thought are not erased. This is a feminism which not only makes space for an orientation to the world that emerges from women's bodies specifically, but is also a world in which sexual difference provides a mechanism for the endless proliferation of difference. Feminism is not for Grosz a critical project but a creative one: '*At its best*', she writes, 'feminist theory is about the invention of the new: new practices, new positions, new projects, new techniques, new values' (83). In this way, Grosz can be seen as a utopian thinker who, like Deleuze, sees a world full of possibility, excess and joy.

## Notes

1. We would like to thank members of our Nonhuman Turn reading group who discussed 'The Nature of Sexual Difference' with us: Katrina Schlunke, Briohny Walker and Susanne Ferwarda. In particular, we'd like to thank Katrina Schlunke, who made us think more critically about the queer potential of the sexual fluids of barnacles.
2. For Grosz this is both the production of difference through reproduction which produces an infinite range of new bodies, but also the difference that manifests through the excesses of sexual selection (the differences that would make individuals desirable). Irigaray is very clear that sexual difference is productive of difference but that this should not be tied to ideas of reproduction and fecundity. 'Sexual difference', she writes, 'would constitute the horizon of worlds more fecund than any known to date – at least in the West – and without reducing fecundity to the reproduction of bodies and flesh. For loving partners this would be a fecundity of birth and regeneration, but also the production of a new age of thought, art, poetry, and language: the creation of a new *poetics*' (2005: 7).
3. Grosz picks up this concept in her 1993 *Topoi* article, an early and important assessment of Deleuze and Guattari's potential use for feminist theory.
4. While this image of sexual selection privileges maleness by attributing to it activity, whilst relegating females to the passive role, Darwin does allow females a degree of agency; he writes: 'The female, though comparatively passive, generally exerts some choice and accepts one male in preference to others. Or she may accept . . . not the male which is the most attractive to her, but the one

which is least distasteful. The exertion of some choice on the part of the female seems almost as general a law as the eagerness of the male' (Darwin 1981: 273, quoted in Grosz 2008: 30). Grosz argues that such an allowance, albeit slim, is indicative of the fact that Darwin 'was less sexist and heterocentric than his contemporaries and especially his neo-Darwinian followers' (2008: 30), although how this indicates a relative lack of heterocentricity, she does not unpack.

5. See, for instance, Donna J. Haraway's 1991 seminal critique of studies into the socio-sexual lives of chimpanzees in *Simians, Cyborgs, and Women: The Reinvention of Nature*.
6. For Grosz, same-sex attraction is also about the specificity of sexual difference because it implies sexual desire for a particular type of body (2011: 131). She writes, 'I cannot see how an understanding of sexuality, sexual pleasure, desire, and identity can be developed which doesn't discern, as part of its very operations, the relative values of and attraction to the particularities of male and female bodies, organs, and activities' (108).
7. Darwin himself would not have explained sexuality in such teleological terms. This reading of Darwin is a product of contemporary, ideologically inflected, heteropatriarchal readings of his work. Thanks to Tano S. Posteraro for reminding us of this point.

## References


Bailey, N. W. and M. Zuk (2009), 'Same-Sex Sexual Behaviour and Evolution', *Trends in Ecology and Evolution* 24: 8, pp. 439–46.

Colebrook, C. (2012), 'Sexual Indifference', *Telemorphosis: Theory in the Era of Climate Change, Vol. 1*, ed. T. Cohen, Michigan: Open Humanities Press.

Darwin, C. (1981), *The Descent of Man and Selection in Relation to Sex*, Princeton: Princeton University Press.

Darwin, C. (1998), *On the Origin of Species by Means of Natural Selection, or the Preservation of Favoured Races in the Struggle for Life*, New York: Modern Library.

Darwin, C. (2010), *A Monograph of the Sub-Class Cirripedia, With Figures of all the Species*, London: FQ.

Deleuze, G. (2001), *Difference and Repetition*, trans. P. Patton, New York: Continuum.

Deleuze, G. and F. Guattari (1994), *What is Philosophy?*, trans. H. Tomlinson and G. Burchell, New York: Columbia University Press.

Deleuze, G. and F. Guattari (2004), *A Thousand Plateaus: Capitalism and Schizophrenia*, trans. B. Massumi, New York: Continuum.

Deleuze, G. and F. Guattari (2005), *Anti-Oedipus: Capitalism and Schizophrenia*, trans. R. Hurley, M. Seem and H. R. Lane, Minneapolis: University of Minnesota Press.

Eckstrand, N. (2014), 'Deleuze, Darwin and the Categorisation of Life', *Deleuze Studies* 8: 4, pp. 415–44.

Grosz, E. (1993), '"A Thousand Tiny Sexes": Feminism and Rhizomatics', *Topoi* 12, pp. 167–79.

Grosz, E. (1994), *Volatile Bodies: Toward a Corporeal Feminism*, Bloomington: Indiana University Press.

Grosz, E. (2008), *Chaos, Territory, Art: Deleuze and the Framing of the Earth*, New York: Columbia University Press.

Grosz, E. (2011), *Becoming Undone: Darwinian Reflections on Life, Politics, and Art*, Durham, NC: Duke University Press.

Grosz, E. (2012), 'The Nature of Sexual Difference: Irigaray and Darwin', *Angelaki* 17: 2, pp. 69–93.

Haraway, D. J. (1991), *Simians, Cyborgs, and Women: The Reinvention of Nature*. New York: Routledge.

Irigaray, L. (2005), *An Ethics of Sexual Difference*, trans. C. Burke and G. C. Gill, New York: Continuum.

Parisi, L. (2010), 'Event and Evolution', *The Southern Journal of Philosophy* 48, pp. 147–64.

Pulkkinen, T. (2017), 'The Role of Darwin in Elizabeth Grosz's Deleuzian Feminist Theory: Sexual Difference, Ontology, and Intervention', *Hypatia* 32: 2, pp. 279–95.

Saldanha, A. (2013), 'Bastard and Mixed-Blood Are the True Names of Race', *Deleuze and Race*, ed. A. Saldanha, Edinburgh: Edinburgh University Press, pp. 1–34.

Weinstein, J. (2010), 'A Requiem to Sexual Difference: A Response to Luciana Parisi's "Event and Evolution"', *The Southern Journal of Philosophy* 48, pp. 165–87.

Chapter 4

# Deleuze, Developmental Systems Theory and the Philosophy of Nature

*Michael James Bennett*

During the final fifteen years of the twentieth century, biology – or at least the philosophy of biology – was shaken up by a perspective that came to be known as 'Developmental Systems Theory'. DST represented a reaction against the perceived hegemony of 'genecentric' conceptions of organismic development, according to which ontogeny was to be explained in terms of the activation of commands or instructions coded in the genome, where the information that determines an organism's form and specific traits pre-exists relatively independent of nongenetic factors. In her watershed study *The Ontogeny of Information: Developmental Systems and Evolution*, Susan Oyama described such genecentrism as an updated version of biological preformationism. And, although in that work she distances her perspective from classical epigenesis with its baggage of metaphysical vitalism (2000a: 3), in the introduction to *Cycles of Contingency* (2001), the most comprehensive collection of DST papers, she and two other advocates of the perspective, the evolutionary psychologist Russell D. Gray and the philosopher Paul E. Griffiths, claim that it is a 'thoroughly epigenetic account of development'. In this context the phrase implies that one organismic lifecycle is not straightforwardly a product of its genes or their simple 'activation' or 'triggering', but of a suite of complex interactions between organism and environment and within the organism itself (Oyama et al. 2001: 4).

More recently, some readers of Deleuze have claimed there are convergences between DST and the latter's work, particularly in *A Thousand Plateaus*. This chapter will consider the similarities and differences between the two perspectives and what DST can actually illuminate about Deleuze's views. It argues that the convergence in question is not straightforward, simply a matter of the substantive positions to which the two perspectives are committed. Deleuze and Guattari would probably not entirely agree with the anti-genecentrism of DST, and, although the two perspectives may share an antipreformationist streak, comparing Deleuzo-Guattarian biophilosophy and DST doesn't help us understand all the positions to which the former is committed. That said, I will argue that the comparison can help us understand something about Deleuze and Guattari in an indirect sense, to the extent that one of the controversies into which DST has waded has to do with the demarcation problem or division of labour between science and philosophy. Critics have suggested that, despite what its advocates claim, DST either makes no difference to practising biologists, or its commitment to holism is at odds with the simplifying assumptions necessary to scientific practice. Peter Godfrey-Smith in particular has wondered whether DST is best understood as a scientific programme, a philosophy of nature, or both. Following this line of thinking, I will argue, makes at least partial sense of Deleuze's occasionally disparaging remarks about the philosophical use of metaphors throughout his work, and of the distinction between science and philosophy in Deleuze and Guattari's final collaboration, *What is Philosophy?*

## Developmental Systems and Blocks of Becoming

Defenders of DST claim that it represents a wholesale attempt to get beyond the simplistic oppositions that have fascinated the public but stymied research: 'nature or nurture, genes or environment, biology or culture. Developmental systems theory . . . is an attempt to do biology without these dichotomies' (Oyama et al. 2001: 1). Take, on the one hand, the distinction between gene and organism. In contrast to most versions of the so-called 'interactionist consensus', which aim to apportion the causal contributions of genes and other factors, DSTers see no reason a priori to assign such prominence to the gene's role in development. The distinction between genes and everything else is only one among many possible distinctions, 'possibly helpful for some purposes, much less so for others', and certainly not categorical (2). In fact, according to the thesis of *The Ontogeny of Information*, the 'blueprint' or 'program' for development in the chromosomes itself emerges from the transformation of organised structures in the ontogenetic context, that is, from interactions among cells within an embryo, and between the embryo and its outside. Moreover, the developmental process producing information itself doesn't require a central governing source, which Oyama compares to an Aristotelian entelechy and satirises as the 'homunculoid' or 'animistic' gene (2000a: 4, 13–14). Talk of genes 'initiating' development only makes sense if you arbitrarily decide to begin analysis at a certain point in an ongoing process (40). Oyama's general point is that genetic information doesn't exist outside the flux of physical causes, and so invoking it as an ultimate explanation has misleading, even transcendental, resonances.

On the other hand, there's the distinction between organism and environment: 'Just as there are no preexisting representations or instructions that shape organisms from within, there are no preexisting niches or environmental problems that shape populations from without' (Oyama et al. 2001: 6). Both the internal dichotomy of organism and gene and the external dichotomy of organism of environment no longer stand up to the developmental systems theorist's scrutiny. DSTers therefore emphasise the importance and pervasiveness of 'niche construction', in which organisms shape their environments and thereby modify the selection pressures to which they are exposed (Griffiths and Gray 1994; Jablonka 2001; Laland et al. 2001). Earthworms, for example, transform the soil in which they live by consuming dead organic matter and excreting nutrients and by mixing or turning over the topsoil and increasing its porosity. Other organisms even transform other organisms – as when the parasitic wasps that famously disturbed Darwin inject their offspring into the bodies of other animals and convert them into 'food factories for their young' (Gould 1982), and, in general, in cases of obligate symbioses, like leafcutter ants and their fungal cultivars (Hölldobler and Wilson 1990: 596–608; cf. Sterelny 2001: 334).

Deleuzians have picked up on both the critique of transcendent genes and the emphasis on symbioses. Protevi, for example, notes that when Deleuze distinguishes between the 'differentiated matter' of the genome and its 'actualization' effected by the cytoplasm (1994: 251), he is 'prefiguring a move' in DST 'away from a self-identical and transcendent genetic program to a differential network of genetic and epigenetic factors' (2012: 246). Likewise, Deleuze and Guattari's famous account of the role of symbioses in evolution – such as the 'block of becoming that snaps up the wasp and the orchid' (1987: 238) – reminds him of the breakdown of the dichotomy between organism and environment in DST. DSTers refer to both of these phenomena to support their view of 'extended inheritance' (Oyama et al. 2001: 3–4), according to which much more than DNA is 'passed on' – including epigenetic factors like chemical gradients in the cytoplasm, endosymbionts and even constructed niches or ecologies (Laland et al. 2001; Griffiths and Gray 2001). Protevi uses the Deleuzo-Guattarian concept and example to express the DSTer's conviction that 'It's the "becoming" of the organism-in-its-niche that needs to be thought as the unit of evolution (e.g., the "wasp-orchid")' (2012: 253). On this view, *Difference and Repetition* speaks to the collapse of the internal dichotomy of gene and organism and *A Thousand Plateaus* to the external one, of organism and environment.

Other readers have noted related connections. Keith Ansell-Pearson originally perceived the link between Deleuze and Guattari's 'unnatural nuptials' in the 'Becoming-Animal' plateau and Lynn Margulis's work on endosymbiosis as a major source of variation for natural selection – a crucial predecessor and theoretical touchstone of DST (Ansell-Pearson 1999: 144, 157, 165–6; cf. Oyama 2000a: 142; Griffiths and Gray 2001: 210). He also argues that Deleuze and Guattari stand in an 'important relation' to a 'movement within contemporary biology . . . in which the emphasis is on the "co-evolution" of organism and environment' and which aims 'to dissolve the opposition . . . and to dereify it as a subject-object split' (1999: 146), which quasi-Hegelian terms are reminiscent of Richard Lewontin's 'dialectical biology' (Levins and Lewontin 1985), another crucial precursor to DST (Oyama 2001). Ann Burlein (2005) has compared Deleuze's Bergson-inspired distinction between the actualisation of the virtual and critique of the possible with Oyama's ontogeny of information.

While there are a number of conceptual commonalities, we should be wary of concluding too much on the basis of them. In *A Thousand Plateaus*, Deleuze and Guattari are more 'genecentric' than a DSTer would like. John Marks has thoroughly documented the influence of François Jacob and Jacques Monod on the text, two major biologists not only committed to the reductionist 'Central Dogma' of molecular biology – that genetic information flows in one direction, from DNA to proteins, which forecloses the possibility of epigenetic or environmental influence on genes beyond 'activation' or 'triggering' – but who also developed the 'operon' model of the gene that 'added a crucial dimension to the informational understanding of life that had begun to develop in the 1940s' (Marks 2006: 86). They appear in Oyama's *Ontogeny of Information* as prominent representatives of the kind of thinking and writing about genes and development to which she objects. Probably because of this influence of Jacob and Monod, Deleuze and Guattari insist on the 'autonomous and independent line' of deoxyribonucleic code in a way that actually tends to be at odds with their interest in organism-level syntheses and 'blocks of becoming' (1987: 59; cf. Allen, this volume). Mark Hansen (2000) has described Deleuze and Guattari as being staunchly committed to molecular reductionism. That would explain why they insist, for example, that the 'becoming-wasp of the orchid and . . . becoming-orchid of the wasp' does not occur simply at the 'level of the strata' (that is, the stratified organisms), but as a 'surplus value of code' (1987: 10), for which the model is lateral gene transfers effected by bacteria (53), a phenomenon described by Jacob and consistent with his genecentrism (Marks 2006: 94). And in *A Thousand Plateaus* Deleuze and Guattari are fittingly hostile to the 'organism' as an image of thought and declare it the 'enemy' of the Body without Organs (1987: 158; cf. Protevi 2001). We are, it seems, a long way from the DST perspective which takes not the gene but the '"becoming" of the organism-in-its-niche . . . as the unit of evolution' (Protevi 2012: 253).

In the most straightforward sense – in terms of the actual commitments of the respective theoretical positions – the parallel Deleuzians detect with DST can only be taken so far. For the rest of this paper I want to consider it in a less straightforward way.

## Research Programmes and Philosophies of Nature

Advocates of DST argue that the perspective is not only consistent with scientific practice, but that it makes better science. For example, many of Oyama's formulations lay the stress on the superior descriptive or explanatory *adequacy* of DST by implying that thinking too informationally or gene-centrically is 'misleading', 'illusionary' or 'incoherent'. Take the following representative passage:

> Because this metaphor of a prior plan seems not only harmless but useful and even necessary . . . it seems difficult to object to it, but I believe it is often misleading. Because we are unaccustomed to thinking in other ways, it invites literal interpretation, and thus the illusion of having given a causal account . . . Once taken literally, the notion of preexisting prescriptive rule commits us to a view of development that is at once incoherent and pernicious. It is incoherent because it is inconsistent with the phenomena it seeks to explain. It is pernicious because it is just this conception of developmental causation (that some development proceeds by genetic rules and some does not) that undergirds the opposition of biological to cultural processes, the mare's nest of biological determinism and the whole nature-nurture complex. (2000a: 12–13)

Griffiths and Gray have made promises on behalf of the scientific usefulness of DST in terms of its *explanatory power*. It will afford biologists 'maximum explanatory power' to construe evolution DST-style as the 'differential replication of total developmental processes or life cycles'

(1994: 278). For example, it could help to explain the role of culture in human and non-human evolution (302, 304).

Despite these foreseen benefits, DST has been criticised for not contributing much to scientific practice because it fails to 'offer anything that aspiring researchers can put to work' (Kitcher 2001: 408). The objection is often coupled with, and perhaps related to, the suggestion that DST is too 'holistic'. As Sterenly et al. put it in a critical essay, the perspective invites questions about how to delimit and individuate developmental systems so that it makes sense to call each of them a 'unit' on its own and not just a part of a larger whole: 'some causal influences are going to be part of the developmental system . . . and others are not, on pain of all developmental systems reducing to one. That would be holism run amok' (1996: 383; cf. Griffiths and Gray 1994; 2001: 209–13). Godfrey-Smith connects the lines of criticism: 'mostly DST does not offer anything that aspiring researchers can put to work, and when it does offer something, what it offers is an insistence on taking every causal factor in biology equally seriously in every investigation, an insistence that would simply shut down scientific activity' (2001a: 283).

In response DSTers have articulated lists of their own research questions: for example, to 'study the longevity and fidelity of extragenetic inheritance' (Gray 2001: 202; cf. Griffiths and Gray 2001: 214–15; Depew and Weber 2001: 240–2). But these lists arguably contain no suggestions that aren't already existing research programmes in genetics or evolutionary developmental biology (Robert et al. 2001: 959). Godfrey-Smith takes another approach in defence of DST. Although he believes that it has at least the potential to steer research, the value of the perspective doesn't stand or fall with this question. That's because scientific practice is not the *only* intellectual project to which DST contributes. His suggestion is, I think, of interest for understanding what it is possible to learn from the connection between Deleuze and DST.

According to Godfrey-Smith, the developmental systems perspective also contributes to what he calls the 'philosophy of nature'. By that he means the respectable project of 'comment[ing] on the overall picture of the natural world that science . . . seem[s] to be giving us' (2001a: 284). Elsewhere he expands on the idea: philosophers of nature

> aim to use scientific work to inform our view of the world, but we do not determine this view using science in its 'raw' form. Instead we take the raw science on a given topic and work out, philosophically, what exactly the work is saying. . . . scientific information generally needs processing before it feeds into discussions of those kinds. (2009: 3)

DST does just this. In addition to proposing research programmes, it helps to interpret ('process') the results of existing programmes philosophically. That means it's both continuous and, crucially, discontinuous with science – unlike self-styled 'naturalistic philosophy', which demands that all good work in the field be 'contained in natural science' (Quine 1969: 83). Such naturalism, Godfrey-Smith says, 'gives up the autonomy of philosophy with respect to its choice of questions' (2001a: 284). In contrast, while 'philosophy of nature' must be consistent with science, its concepts and questions need not be identical to scientific ones – though Godfrey-Smith leaves open the possibility that they *could* be: 'Any degree of overlap is possible' (284).

Charting a middle course, Godfrey-Smith also distinguishes philosophy of nature from *Naturphilosophie*, with its pejorative connotations of armchair philosophising (cf. Quine 1969: 126). Rather than being an a priori propaedeutic, 'Philosophy of nature in my sense comes *after* empirical science' and doesn't foist or project abstract philosophical schemas onto nature without acquaintance with it (2001a: 284). This distancing gesture is perhaps a little ironic, considering that Godfrey-Smith first used the phrase 'philosophy of nature' in an essay on Lewontin's 'dialectical' (that is, Hegelian-Marxist) biology (2001b). Lewontin wasn't a dogmatic Hegelian, but he was a powerful critic of the pervasive analytic method of explanation in science that breaks down every system into its parts and assumes that the important properties for understanding it pre-exist in those parts rather than emerging through their interactions, a view he called 'Cartesian [i.e., mechanistic] reductionism' (Levins and Lewontin 1985: 2–6). Lewontin argues that modern genecentrism in ontogeny and adaptationism in phylogeny are versions of this non-dynamic analytic worldview. To make the point he employs quasi-Hegelian generalisations: 'For Mendel . . . what we now call "genes" were the subjects and the organisms the objects of developmental forces. . . . For Darwin . . . the external world with its autonomous properties, was the subject and the organism was, again, the object acted upon' (1983: 273). By refocusing on the ways in which organisms are causes of their own development and constructors of their environments, Lewontin suggests, biologists will be better equipped to recognise the dialectical interpenetration of cause and effect, subject and object in gene-organism and organism-environment relations. A perspective preferable to 'Cartesian reductionism' will grasp that the 'organism is not simply the object of developmental [and environmental] forces, but is the subject of these forces as well' (279). For Godfrey-Smith, such 'dialectical' biology, far from being speculative *Naturphilosophie*, is a contribution to the philosophy of nature in his sense of the term. It 'comes after' biological inquiry and aims to comment on the overall picture that it is delivering. It's to Lewontin's credit that he has contributed both to the intellectual project of science and to the philosophy of nature (Godfrey-Smith 2001b: 265), and DST straddles the same disciplinary boundary (2001a).

But what is the point of doing the philosophy of nature? What, for example, is Lewontin trying to achieve by describing genetics and ecology in terms of 'subjects', 'objects' and 'dialectics'? Godfrey-Smith explains that the philosophy of nature has at least two functions (2001a). First, scientific descriptions must be 'processed' for broad consumption for the same reason that raw and potentially unhealthy food is – public health. For example, talk of 'genes for' traits – no matter how scientifically useful it is or how logically respectable it can be made (e.g., by Sterelny and Kitcher 1988) – tends to encourage a kind of genetic determinism, which no self-respecting scientist would actually endorse (cf. Oyama 2000a: 13). As philosopher of nature, Lewontin relentlessly criticised the vulgarisations of sociobiology, particularly in public forums such as the *Tanner Lectures on Human Values* (1983) and the *Massey Lectures* (1990). DSTers are similarly concerned about the popular uptake of scientific concepts. Griffiths and Gray begin their landmark paper with the critical observation that 'Few scientific ideas are as well embedded in popular culture as the idea that certain features of an organism are genetically determined' (1994: 277). Jason Scott Robert offers a particularly sophisticated version of this defence of DST as 'philosophy of nature' (though he doesn't use this term). He thinks that the DST perspective can help to guard against the proliferation of inferential tricks and 'invalidly inferred' conclusions about the role of genes in development (2004: 127–8). For example, he diagnoses the phenomenon of 'hedgeless hedging' among developmentalists: a strategy of *appearing* to hedge one's bets – e.g., 'betting' that development is a matter of gene action but making a side bet on nongenetic factors just in case that's not true – without really hedging them, or without really expecting to find any explanation other than gene action (2003). Robert hopes that the 'methodological systematism' of DST can expose such arguments and that its sympathisers will be 'vigilant in assessing the inferences drawn' from scientific results (2004: 128).

Such vigilance is *de rigeur* because of the widespread use of metaphors in actual science. Lewontin is fond of quoting Norbert Wiener: 'The price of metaphor is eternal vigilance' (Lewontin 2000: xv). According to Godfrey-Smith, dealing with and making sense of scientific metaphors is therefore one other essential function of the philosophy of nature (2001a:

287–8). It explains why the use of language in philosophy of nature must have, at least potentially, a degree of independence from the scientific usage. Oyama is exemplary when she argues that the problems associated with the notion that biological form in some sense pre-exists its ontogenetic unrolling are in no way diminished by any of the series of metaphors used to express the idea (2000a: 1), and devotes a memorable chapter of *The Ontogeny of Information* (Chapter 5) to deflating each in turn. Metaphors simplify the complexity of natural dynamic systems, which may lead to popular misunderstandings of scientific results or even to scientists producing inferior science.

The trouble with metaphors (the fact that they simplify) is, however, also their greatest virtue. Godfrey-Smith imagines a scenario: 'Suppose one holds that metaphorical and analogical language is essential to science . . . [that] ordinary scientific work requires . . . the constant deployment of a rich set of metaphors' (2001a: 287). For example, it might be no accident that the proliferation of Oyama's nemesis, the locution 'genetic coding for', has corresponded with massive progress in genetics over the past century. One might argue that because of either human cognitive limitations or the intrinsic complexity of the subject matter (or both) normal science in genetics requires 'some simplifying framework or other' and the question is just 'which is the simplifying framework best suited for us' (287). In a nutshell, the suggestion is that the metaphors DSTers criticise are the simplifying strategies normal science finds indispensable. In other words, metaphors like 'genes are codes' function as what Robert (following Wimsatt 2007) calls 'heuristics': 'simplifying strategies to be used in situations of cumbersome investigational complexity' (2003: 976), which make that complexity tractable. Robert summarises: 'Experimental tractability is a core scientific desideratum. It is nice to imagine the world as full of interconnected parts not meaningfully separable from each other; but just try to analyze the world so imagined, and science grinds to a halt' (979). Kitcher may endorse the view suggested by Godfrey-Smith when the former remarks that DST consists primarily of 'critiques of the past misuses of old tools' rather than suggestions for new ones (2001: 408). If metaphors, even misleading ones, were 'tools', then that suggests they were useful.

So does DST demand that we do without *all* metaphors, heuristics and simplifying assumptions in science? That we 'do biology' without them, just as we ought to do it without the convenient but ultimately deceptive dichotomies of nature/nurture, genes/environments, and so on? If so, and if heuristics are crucial for research getting traction, then it sounds as if DST requires an 'extreme holism . . . inconsistent with the practice of ordinary empirical research' that confirms the views of its critics

(Godfrey-Smith 2001a: 283; cf. Sterelny et al. 1996). Oyama does sometimes sound as if she thinks all metaphors and heuristics need to go: 'Though investigators suddenly bereft of the nature-nurture context . . . may wish for a replacement, it may be just our assumption that such large-scale simplifying research guides are possible and necessary that needs to change' (2000a: 153). But she also argues that science needs *better* metaphors, not to be metaphor-free: 'The remedy is surely not a draconian prohibition on simile, hyperbole, and the poetic imagination. Literary license, though, is not license to mislead, and when metaphor is employed in the service of scientific understanding, it should be *accurate* and *helpful* as well as vivid and evocative' (130, my emphasis). Other DSTers follow suit and advocate replacing certain metaphors with more accurate and helpful ones (e.g., Griffiths and Gray 1994: 284–5).

Oyama's invocation of 'accuracy' suggests that 'DST makes an empirical bet that biology would be better off' without misleading metaphors like 'genes are codes for traits' (Godfrey-Smith 2001a: 287). Nevertheless, as Will Wimsatt has articulated forcefully, sometimes the two requirements Oyama mentions – accuracy and helpfulness – are at odds when it comes to heuristics: in fact, even certain false models can produce useful insights (2007: 94–106). Wimsatt actually enumerates *twelve* functions of false models. For example, 'neutral models' in evolutionary biology – models of descent that exclude selection – can be deployed self-consciously as false baseline patterns and compared with experimental data to evaluate the effect of excluded variables (here, selection). Wimsatt's aim is to restrain the philosophers who argue from the widespread use of false models in scientific practice to scientific antirealism, while many scientists embrace a local, modest realism on a heuristic rather than a metaphysical basis. But the distinction between science and philosophy of nature can also respond to this situation. Even supposing that the DSTer's 'bet' is wrong and that simplifying metaphors are good for science, that result applies *only* to the scientific face of DST, not its role as philosophy of nature. While a scientist may have good, pragmatic reasons for using certain heuristics, the philosopher is free to find the falsity of 'false models' intolerable because she is committed to accuracy, not just helpfulness. Godfrey-Smith expresses the thought in terms of literal and nonliteral meaning:

> The philosopher of nature should ask: even if this way of talking has great practical value, what is its real status? Is this a good literal description of [e.g.] what genes are like, or is it a metaphor? Is it perhaps language that is being applied nonmetaphorically and being used to make literally false claims, but that nonetheless have heuristic value? (2001a: 288)

This explanation of the DSTer's attitude toward metaphor makes good sense of Oyama's view, which I quoted earlier, that metaphors like 'prior plan' really become a problem when they're taken literally by both scientists and everyone else (2000a: 13).

In sum, Godfrey-Smith interprets DST as equivocal, as a contribution both to science and to philosophy of nature, and he seems to think that DSTers haven't sufficiently come to grips with this equivocation. This construction of the distinction between science and philosophy may, moreover, have certain implications that can ultimately help us to think about the comparison between DST and Deleuze. One implication could be that the philosopher of nature is committed to a kind of *truth* independent of what happens to be useful for scientists, who may find what is metaphorical and 'literally false' helpful – such as the idea that the genes are a 'blueprint' for development. That is, the 'heuristic value' of potentially 'false models' becomes a problem for DST *qua* philosophy of nature. Another implication has to do with different attitudes toward metaphor characteristic of science and philosophy. It may be that as a scientist Oyama wants to develop better simplifying heuristics (metaphors), but as a philosopher she wonders about the necessity of metaphors as 'large-scale simplifying research guides' in general. The equivocation of DST as a whole explains Oyama's vacillation on metaphor.

## Deleuze Against Metaphor

Deleuze also shares with DST a strong suspicion of metaphor. For example, in the same plateau as the discussion of 'blocks of becoming', he and Guattari criticise two other ways of conceiving the relationship between animals: first, as an imaginary series progressing toward perfection, as in, for example, Aristotle's zoology (cf. Bennett 2017: chap. 6), and second, as a conceptual structure consisting of ordered differences and correspondences between relations, as in Lévi-Strauss's analyses of totemism (Deleuze and Guattari 1987: 233–6; cf. Deleuze 2004: 173 and 2007: 202). Deleuze and Guattari profess to 'believe in the existence of very special becomings-animal' (1987: 237) to which Aristotelianism and structuralism were blind. They mean symbioses like the exemplary wasp-orchid. The difference between imaginary series and conceptual structures, they say, is like the difference between the analogy of proportion and of proportionality, and, more importantly for my purposes here, that shifting from thinking in terms of series to thinking in terms of structures involves replacing 'metamorphoses of the imagination with conceptual metaphors' (1987: 236; cf. Deleuze and Guattari 2003: 35–6). Likewise, in the analysis of order-words earlier in the same work, Deleuze and Guattari say that 'the first determination of language, is not the trope or metaphor but *indirect discourse*. The importance some have accorded metaphor and metonymy proves disastrous for the study of language. Metaphors and metonymies are merely effects' (1987: 76–7). Elsewhere, Deleuze is at pains to indicate that he's not speaking metaphorically (1990: 76; 1994: 25, 181, 190; 1989: 57) and associates metaphor with analogical thinking (1994: 24; 1990: 360 n.5; 1995: 29). Several of Deleuze's commentators have noted that, despite appearances, Deleuze and Guattari are not speaking metaphorically when they talk about 'stratification' in *A Thousand Plateaus* (Bonta and Protevi 2004: 9, 15), nor is Deleuze when he mentions mathematical 'singularities' in *Difference and Repetition* and elsewhere (DeLanda 2011: 18). He may consider 'speaking metaphorically' to imply a kind of veiled Platonism that moves from 'meanings sedimented in language in the order of discovery (instances, copies) to those later discovered but prior in the order of being (forms)' (Bonta and Protevi 2004: 113), or a one-dimensional transfer of sense that 'obliterates . . . generic boundaries for new resemblances' (Martin 2010: 172), a construal of what metaphor involves that Deleuze himself approves (2007: 361).

There is thus something about the function of metaphor in general that Deleuze sees as collapsing distinctions and obliterating part of the rich texture he wants his philosophy to capture. Unlike Oyama, Deleuze doesn't claim to be replacing bad metaphors with better ones, but seems to take a more hard-line approach. Still, some may find it peculiar to believe that a proposition like 'Wasp and orchid . . . form a rhizome' (Deleuze and Guattari: 1987: 10) is not a metaphor. So what's the best way for us to understand Deleuze's distrust? I want to suggest that his views on metaphor become clearer when we consider them in relation, first, to the disciplinary distinction between science and philosophy, which Deleuze discusses in a number of places but most systematically in *What is Philosophy?*, and second, to the idea that *metaphors can be heuristics* suggested by the discussion of DST.

In *What is Philosophy?*, Deleuze and Guattari famously propose their own division of labour between science and philosophy. Both modes of thought are manners of relating to the 'infinite movement', which thought claims 'by right' (1994: 37), or to 'chaos', defined both as the 'infinite speed with which every form taking shape in it vanishes' and as 'a *virtual* containing all possible particles and drawing out all possible forms, which spring up only to disappear immediately, without consistency or reference' (118). Without saying too much about the notion of the 'virtual' here, let it suffice to point to part of the oft-quoted definition: it refers to what is 'real without being actual' (Deleuze 1994: 208–9; 1991: 96–7).

Given this ontological scene setting, the difference between science and philosophy as relationships with infinitely fast, virtual chaos is pretty straightforward: philosophy *retains* the infinity and virtuality of chaos, while science *relinquishes* both for the sake of other aims. Philosophy 'wants to know how to retain infinite speeds while gaining consistency, by *giving the virtual a consistency specific to it*' (Deleuze and Guattari 1994: 118; cf. 42). A philosopher aims to do so by creating concepts, which have their own 'specific infinity' (21) and are 'real without being actual' (22), and, correlatively, by slicing through chaos with a 'plane of immanence or consistency' (118), an 'object of infinite specification so that it seems to be a One-All' (39). Science, in contrast, 'approaches chaos in a completely different, almost opposite way: it relinquishes the infinite, infinite speed, in order to gain *a reference able to actualize the virtual*' (118) – in other words, it approaches chaos in such a way that it is no longer infinite or virtual but finite and actual. Scientists achieve this by creating 'functions' (or 'functives'), rather than concepts, and installing a 'plane of reference', which is 'like a freeze-frame . . . a fantastic slowing down' (118).

The general point seems to be that science, unlike philosophy, has an interest in strategically simplifying complexity. The complexity in question might be the absolutely 'dissipative' quality of chaos, which undoes forms as fast as they appear (Cooper 2002), or the unwieldy infinity of the 'One-All' or Epicurean *pan apeiron* (Bennett 2017: chap. 9). Philosophy, however, strives to do justice to just these impossibly complex realities. To redeploy Oyama's terms, the 'infinite speed' that philosophy aims to retain is not *helpful* to scientific endeavour. On the other hand, even though science is interested in the actual, to which it makes reference, it is not as *adequate* to the (virtual) real as philosophy is.

The trouble with this suggestion is that adequacy (or adequation) is a property of representations and has a longstanding affiliation with truth, while Deleuze and Guattari flatly deny that philosophy is 'inspired by truth' (1994: 82). In a related way, unlike scientific functions and propositions, which refer to bodies and actual states of affairs, philosophical concepts express events (22, 33, 133). Deleuze's views about truth and reference are complex and distinctive (Bennett 2017: chap. 2), but I hope it is not too much of a distortion to suggest that by saying that philosophy 'retains' something of infinitely fast, virtual chaos Deleuze and Guattari imply that philosophical concepts and virtual events stand in a certain relationship of commensurability that is reminiscent of the adequation traditionally involved in truth (Allen 1993). Paul Patton (1996) has argued that, even though they say that philosophy doesn't make reference and isn't interested in truth, Deleuze and Guattari nevertheless vacillate on its 'cognitive status' – asserting both that philosophy 'does not consist in knowing and is not inspired by truth' (1994: 82), and that a philosophical concept is 'obviously knowledge – but knowledge of itself, and what it knows is the pure event' (33). Perhaps there's a similar vacillation going on here in terms of the dissociation of philosophy and truth and the simultaneous implication that philosophy is more adequate than science to the (virtual) real.

As Jeff Bell notes, the sharp distinction between philosophy and science is 'One of the more controversial moves in *What is Philosophy?*' (2016: 158). Those accustomed to Deleuzo-Guattarian affirmations of productive connections and symbiotic becomings may feel betrayed by the unambiguous, even neo-classical division of theoretical labour (Stengers 2005). Some may doubt that the division is really clear-cut at all, even in the text of the book itself, which continues Deleuze's lifelong practice of appropriating scientific notions for his own philosophical ends (Schmidgen 2015). Perhaps, as Todd May has argued, Deleuze's appropriations can best be understood as attempts to '"speed up" scientific viewpoints by offering them an ontological perspective that . . . brings them into contact with pure difference' (2005: 254). This way, 'functives . . . double as concepts' (254), or in other words, one and the same way of thinking has two aspects, two sides: scientific and philosophical. This suggestion dovetails nicely with Godfrey-Smith's claim that scientists sometimes wear two hats – as practitioners and as philosophers of nature. Lewontin and the DSTers, for example, contribute to two projects simultaneously, and it would be a kind of category error to criticise their contributions to one intellectual endeavour for failing to live up to the demands of the other – say, by faulting a philosophy of nature for not offering feasible suggestions for research or by demanding that scientists abandon their simplifying strategies like philosophers do.

In this light, my discussion of DST and related issues can help to interpret Deleuze and Guattari's controversial distinction. Above I claimed that their view is that science, as a practice, relies on installing simplifying frameworks ('planes of reference') to deal with ontological complexity ('infinite speed', 'chaos'). Another way to say that is: according to Deleuze and Guattari, science is defined by and coextensive with the use of heuristics. This makes good sense of the suggestion that science 'relinquishes' something. It aims for experimental tractability by bracketing some 'cumbersome investigational complexity'. Philosophy, which need not adopt any such simplifying assumptions in order to proceed, may therefore actually be more adequate to the real – provided the real is appropriately qualified and not understood in actualist terms. In the discussion of DST I described how the specific metaphors endemic to genetics and biology are (by hypothesis at least) good examples of such heuristics. For example, genes are instructions activated or triggered by nongenetic factors. It may well be that this is an *inadequate* or even *false* statement, but it's still *helpful* for generating scientific knowledge. That is the bullet that, as I understand them, Wimsatt and Godfrey-Smith are willing to bite. I suspect Deleuze and Guattari would bite it too, in favour of an antirealist science and a realist philosophy of the virtual.

In fact, the convergence of Deleuze and Guattari with what is at stake philosophically in DST is even more revealing. It also explains, in part at least, their aversion to metaphor. They claim that 'Philosophy proceeds with a plane of immanence or consistency; science with a plane of reference' (1994: 118). They had already written in *A Thousand Plateaus*, right after an allusion to the wasp-orchid, that 'The plane of consistency is the abolition of all metaphor; all that consists is Real' (1987: 69). From the point of view of the interpretation of *What is Philosophy?* that I have urged here, the abolition of metaphor means getting rid of heuristics. Those are for scientists, not philosophers, and Deleuze describes even his most experimental work as 'just plain old philosophy' (2007: 176).

Admittedly, this is somewhat (but not entirely) at odds with Deleuze's own ways of speaking. He does talk about metaphor in the context of describing the difference between science and philosophy, but in a different way to what I've just sketched. For example in *Cinema II*, Deleuze remarks, 'Of course we realize the dangers of citing scientific propositions outside their own sphere. It is the danger of arbitrary metaphor or forced application' (1989: 129). Likewise, in an interview Deleuze responds to the question of whether *A Thousand Plateaus* uses scientific concepts in a metaphorical sense:

> I'd like to reply by saying there are two sorts of scientific notions . . . There are notions that are exact in nature, quantitative, defined by equations, and whose very meaning lies in their exactness: a philosopher or writer can use these only metaphorically, and that's quite wrong, because they belong to exact science. But there are also essentially inexact yet completely rigorous notions that scientists can't do without, which belong equally to scientists, philosophers, and artists. (1995: 29)

In this passage we find both an anticipation of the notion of a scientific 'functive' and a certain way of speaking about metaphor, consistent with *Cinema II*. Here 'metaphor' is what's *defective* in a philosophical use of certain scientific concepts, not (as I have suggested) what's necessary and endemic to scientific practice. In both cases, however, it's something that philosophy would be better off without.

## Holism

I want to conclude with some reflections on the theme of holism. If, as Godfrey-Smith claims, developmental systems theory has a foot in two camps, science and philosophy of nature, then the heuristic helpfulness of misleading metaphors and simplifications will be a problem for DST to the extent that it's a philosophy of nature committed to a kind of realistic adequacy, but not insofar as it's a contribution to scientific practice. In *What is Philosophy?* Deleuze and Guattari are unambiguous about the distinction between science and philosophy and clear-sighted about their diverging aims, in a gesture that Godfrey-Smith might think DST can learn from. In contrast, the charge of extreme, unwieldy holism will only be a problem for DST to the extent that it claims to be a biological research programme, *not* to the extent that it's a philosophy of nature (Godfrey-Smith 2001a: 289). Thus, with their philosopher of nature hats on, DSTers can be as holistic as they like and aim to do justice to, as Oyama puts it, 'all influences on development, at all levels of analysis' (2000b: 72). The same applies to Deleuze and Guattari. They aren't aiming to influence scientific practice, and so their 'philosophy of nature' – a label that, as it happens, they embraced at least once (Deleuze 1995: 155) – defines philosophy in relation to a certain kind of whole. The plane of immanence that philosophy characteristically installs is a 'One-All'.

There are a few grounds for thinking that Deleuze's thought involves a kind of holism. Ansell-Pearson (2002, 2007) has considered Deleuze's debt to Bergson's conception of philosophy as the 'effort to dissolve again into the Whole' (Bergson 1911: 191). May has argued that the rhizome is the best image for what he calls 'contingent holism', the antifoundationalist antireductionism he thinks is common to Nancy, Derrida, Levinas and Deleuze (1997: 59). But whether or not Deleuze's philosophy of nature is fairly described as 'holistic' – a fraught term to be sure – the point is just that it *could* be without risking the kind of embarrassment DSTers want to avoid. Holism is a problem to the extent that they want their statements to be useful for practising scientists, but Deleuze doesn't worry about that in the same way. He's open to the possibility and remarks that it's 'not impossible for a philosopher to create concepts that can be used in science' (1995: 30), but just in the same way that Godfrey-Smith describes how philosophy of nature *may*, but certainly *need not*, feed back into science (2009: 4). Deleuze and Guattari do suggest that philosophical concepts and scientific functions 'necessarily intersect' when they reach their 'full maturity' (1994: 161; cf. Stengers 2005). But, like Godfrey-Smith, they're aiming to disambiguate. While science and philosophy may ultimately converge, it is necessary to recognise the independence of the two tasks: 'it is always unfortunate when scientists do philosophy without really philosophical means or when philosophers do science without real scientific means (we do not claim to have been doing this)' (Deleuze and Guattari 1994: 161). Deleuze and Guattari are self-aware; they aren't doing defective science but philosophy of nature.

Despite Godfrey-Smith's suggestion that DST is really two intellectual projects wrapped into one, its supporters haven't embraced the 'holism' to which they may be entitled as philosophers of nature and have continued to salvage parts of DST as a research programme (Griffiths and Gray 1997; Barberousse 2010; Shea 2011). Possibly that's because DSTers and their fellow travellers are particularly wary of a certain Hegelian brand of holism. Dialectical biology, for example, may be tempted toward a conception of the organism-environment system as a whole, taking 'its parts – the organism on the one hand and the environment on the other – to exist only in relation to each other' (Godfrey-Smith 2001b: 258). In other words, this particular philosophy of nature seems to make wholes prior to their parts and to characterise these constitutive parts in terms of 'relations of interiority', where 'the component parts are constituted by the very relations they have to other parts in the whole' and 'A part detached . . . ceases to be what it is' (DeLanda 2006: 9). Godfrey-Smith concludes that Lewontin's commitment to pluralism and aversion to oversimplification help him to resist this bad holism (2001b). In a related meditation on language, Oyama concedes that terms like 'system' may seem to imply such a logic of interiority, but that she 'mistrusts' the privileging of wholes over parts, because it encourages 'exaggerated confidence' in certain outcomes inconsistent with the contingency that DST lionises in the face of genetic determinism (2001: 185). She asks, 'Must *system* be constitutively contaminated by this brand of holism?' and her answer seems to be 'no'. Nevertheless, Oyama appreciates the irony that 'The concept that some find useful in undoing the internalism of traditional notions of development has, for others, its own internalist aura' (187).

If it's fair to call Deleuze's philosophy of nature holistic, then it also strives to avoid being holistic in this 'internalist' sense. As DeLanda explains, that's one aspect of what it means to declare the organism the 'enemy'. Deleuzo-Guattarian assemblages are characterised by their 'relations of exteriority' not interiority. Their model is not the organism but symbioses: matters of contingent obligation rather than logical necessity (DeLanda 2006: 10–11). Strangely, Deleuze seems to be attracted both to holism and to the notion of external relations, that is, relations external to their terms, which historically entailed a rejection of the former – either in the form of a purely negative statement of pluralism (William James) or a dualist ontology of terms and relations (Bertrand Russell) (Madelrieux 2014). Like Lewontin, Deleuze is torn between affirming pluralism and holism. It may be that, as Madelrieux suggests, Deleuze's ontology, with its many dichotomies (event and body, intensive and extensive, virtual and actual, transcendental and empirical) should be understood as a version of Russellian dualism. Yet Deleuze and Guattari sound like Oyama when they remark that 'mental correctives are necessary to undo the dualisms we had no wish to construct but through which we pass' and suggest that they deploy dualisms tactically, 'invok[ing] one dualism only in order to challenge another'. By this method they arrive at the rhizomatic 'magic formula . . . PLURALISM = MONISM' (1987: 20), which all suggests they are quite self-aware of the tension between holism and external relations as well.

For the most part developmental systems theory refuses holism as pernicious to science, which must employ heuristics or simplifying assumptions – paradigmatically, reductive models that are ultimately inadequate or strictly speaking false. It also rejects internalist holism as a philosophy of nature. Oyama's discussion of the matter, however, implies that there may be other 'brands of holism' available to philosophers of nature. Possibly Deleuze and Guattari endorse a non-internalist one. If so, they are perfectly aware that it's a philosophical approximation of infinite complexity in nature, not a contribution to scientific practice.

## References

Allen, B. (1993), *Truth in Philosophy*, Cambridge, MA: Harvard University Press.

Ansell-Pearson, K. (1999), *Germinal Life: The Difference and Repetition of Deleuze*, New York: Routledge.

Ansell-Pearson, K. (2002), *Philosophy and the Adventure of the Virtual: Bergson and the Time of Life*, New York: Routledge.

Ansell-Pearson, K. (2007), 'Beyond the Human Condition: An Introduction to Deleuze's Lecture Course', *SubStance* 36: 3, pp. 57–71.

Barberousse, A. (2010), 'The Role of Self-Organization in Developmental Systems Theory and the Neo-Darwinian Theory of Evolution', *Biological Theory* 5: 3, pp. 202–5.

Bell, J. A. (2016), *Deleuze and Guattari's* What is Philosophy? *A Critical Introduction and Guide*, Edinburgh: Edinburgh University Press.

Bennett, M. J. (2017), *Deleuze and Ancient Greek Physics: The Image of Nature*, London: Bloomsbury.

Bergson, H. (1911), *Creative Evolution*, trans. A. Mitchell, New York: Henry Holt.

Bonta, M. and J. Protevi (2004), *Deleuze and Geophilosophy: A Guide and Glossary*, Edinburgh: Edinburgh University Press.

Burlein, A. (2005), 'The Productive Power of Ambiguity: Rethinking Homosexuality through the Virtual and Developmental Systems Theory', *Hypatia* 20: 1, pp. 21–53.

Cooper, M. (2002), 'Vitesses de l'image, puissances de la pensée: la philosophie épicurienne revue par Deleuze et Guattari', *French Studies* 61: 1, pp. 45–60.

DeLanda, M. (2006), *A New Philosophy of Society: Assemblage Theory and Social Complexity*, New York: Continuum.

DeLanda, M. (2011), *Philosophy and Simulation: The Emergence of Synthetic Reason*, New York: Continuum.

Deleuze, G. (1989), *Cinema II: The Time-Image*, trans. H. Tomlinson and B. Habberjam, Minneapolis: University of Minnesota Press.

Deleuze, G. (1990), *The Logic of Sense*, trans. M. Lester with C. Stivale, New York: Columbia University Press.

Deleuze, G. (1991), *Bergsonism*, trans. H. Tomlinson and B. Habberjam, New York: Zone Books.

Deleuze, G. (1994), *Difference and Repetition*, trans. P. Patton, New York: Columbia University Press.

Deleuze, G. (1995), *Negotiations: 1972–1990*, trans. M. Joughin, New York: Columbia University Press.

Deleuze, G. (2004), *Desert Islands and Other Texts (1953–1974)*, ed. D. Lapoujade, trans. M. Taormina, Los Angeles: Semiotext(e) Foreign Agents Series.

Deleuze, G. (2007), *Two Regimes of Madness: Texts and Interviews 1975–1995*, trans. A. Hodges and M. Taormina, Cambridge, MA: MIT Press.

Deleuze, G. and F. Guattari (1987), *A Thousand Plateaus: Capitalism and Schizophrenia*, trans. B. Massumi, Minneapolis: University of Minnesota Press.

Deleuze, G. and F. Guattari (1994), *What is Philosophy?*, trans. H. Tomlinson and G. Burchell, New York: Columbia University Press.

Deleuze, G. and F. Guattari (2003), *Kafka: Toward a Minor Literature*, trans. D. Polan, Minneapolis: University of Minnesota Press.

Godfrey-Smith, P. (2001a), 'On the Status and Explanatory Structure of Developmental Systems Theory', *Cycles of Contingency: Developmental Systems and*

*Evolution*, ed. S. Oyama, P. E. Griffiths and R. D. Gray, Cambridge, MA: MIT Press, pp. 283–97.

Godfrey-Smith, P. (2001b), 'Organism, Environment, and Dialectics', *Thinking About Evolution: Historical, Philosophical, and Political Perspectives*, ed. R. S. Singh et al., New York: Cambridge University Press, pp. 253–66.

Godfrey-Smith, P. (2009), *Darwinian Populations and Natural Selection*, New York: Oxford University Press.

Gould, S. J. (1982), 'Nonmoral Nature', *Natural History* 91: 2, pp. 19–26.

Gray, R. (2001), 'Selfish Genes or Developmental Systems?', *Thinking About Evolution: Historical, Philosophical, and Political Perspectives*, ed. R. S. Singh et al., New York: Cambridge University Press, pp. 184–207.

Griffiths, P. E. and R. D. Gray (1994), 'Developmental Systems and Evolutionary Explanation', *Journal of Philosophy* 91, pp. 277–304.

Griffiths, P. E. and R. D. Gray (1997), 'Replicator II: Judgment Day', *Biology and Philosophy* 12: 4, pp. 471–92.

Griffiths, P. E. and R. D. Gray (2001), 'Darwinism and Developmental Systems', *Cycles of Contingency: Developmental Systems and Evolution*, ed. S. Oyama, P. E. Griffiths and R. D. Gray, Cambridge, MA: MIT Press, pp. 195–218.

Hansen, M. (2000), 'Becoming as Creative Involution? Contextualizing Deleuze and Guattari's Biophilosophy', *Postmodern Culture* 11: 1, <http://pmc.iath.virginia.edu/text-only/issue.900/11.1hansen.txt> (last accessed 30 July 2018).

Hölldobler, B. and E. O. Wilson (1990), *The Ants*, Cambridge, MA: Harvard University Press.

Jablonka, E. (2001), 'The Systems of Inheritance', *Cycles of Contingency: Developmental Systems and Evolution*, ed. S. Oyama, P. E. Griffiths and R. D. Gray, Cambridge, MA: MIT Press, pp. 99–116.

Kitcher, P. (2001), 'Battling the Undead: How (and How Not) to Resist Genetic Determinism', *Thinking About Evolution: Historical, Philosophical, and Political Perspectives*, ed. R. S. Singh et al., New York: Cambridge University Press, pp. 396–414.

Laland, K. N., F. J. Odling-Smee and M. W. Feldman (2001), 'Niche Construction, Ecological Inheritance, and Cycles of Contingency in Evolution', *Cycles of Contingency: Developmental Systems and Evolution*, ed. S. Oyama, P. E. Griffiths and R. D. Gray, Cambridge, MA: MIT Press, pp. 117–26.

Levins, R. and R. Lewontin (1985), *The Dialectical Biologist*, Cambridge, MA: Harvard University Press.

Lewontin, R. C. (1983), 'Gene, Organism and Environment', *Evolution: From Molecules to Men*, ed. D. S. Bendall, Cambridge: Cambridge University Press, pp. 273–85.

Lewontin, R. C. (2000), 'Foreword', S. Oyama, *The Ontogeny of Information: Developmental Systems and Evolution*, 2nd edn, Durham, NC: Duke University Press, pp. vii–xv.

Madelrieux, S. (2014), 'Pluralism without Pragmatism: Deleuze and the Ambiguities of the French Reception of James', *Deleuze and Pragmatism*, ed. S. Bowden, S. Bignall and P. Patton, New York: Routledge, pp. 89–104.

Marks, J. (2006), 'Molecular Biology in the Work of Deleuze and Guattari', *Paragraph* 29: 2, pp. 81–97.

Martin, J-C. (2010), *Variations: The Philosophy of Gilles Deleuze*, trans. C. V. Boundas and S. Dyrkton, Edinburgh: Edinburgh University Press.

May, T. (1997), *Reconsidering Difference: Nancy, Derrida, Levinas, Deleuze*, University Park: Pennsylvania State University Press.

May, T. (2005), 'Gilles Deleuze, Difference, and Science', *Continental Philosophy of Science*, ed. Gary Gutting, Oxford: Blackwell, pp. 239–54.

Oyama, S. (2000a), *The Ontogeny of Information: Developmental Systems and Evolution*, 2nd edn, Durham, NC: Duke University Press.

Oyama, S. (2000b), *Evolution's Eye: A Systems View of the Biology-Culture Divide*, Durham, NC: Duke University Press.

Oyama, S. (2001), 'Terms in Tension: What Do You Do When All the Good Terms Are Taken?', *Cycles of Contingency: Developmental Systems and Evolution*, ed. S. Oyama, P. E. Griffiths and R. D. Gray, Cambridge, MA: MIT Press, pp. 177–93.

Oyama, S., P. E. Griffiths and R. D. Gray (2001), 'Introduction: What is Developmental Systems Theory?', *Cycles of Contingency: Developmental Systems and Evolution*, ed. S. Oyama, S., P. E. Griffiths and R. D. Gray, Cambridge, MA: MIT Press, pp. 1–11.

Patton, P. (1996), 'Concept and Event', *Man and World* 29, pp. 315–26.

Protevi, J. (2001), 'The Organism as the Judgment of God: Aristotle, Kant and Deleuze on Nature (that is, on biology, theology and politics)', *Deleuze and Religion*, ed. Mary Bryden, New York: Routledge, pp. 30–41.

Protevi, J. (2012), 'Deleuze and Life', *The Cambridge Companion to Deleuze*, ed. D. W. Smith and H. Somers-Hall, New York: Cambridge University Press, pp. 239–64.

Quine, W. V. O. (1969), *Ontological Relativity and Other Essays*, New York: Columbia University Press.

Robert, J. S. (2003), 'Constant Factors and Hedgeless Hedges: On Heuristics and Biases in Biological Research', *Philosophy of Science* 70: 5, pp. 975–88.

Robert, J. S. (2004), *Embryology, Epigenesis, and Evolution: Taking Development Seriously*, New York: Cambridge University Press.

Robert, J. S., B. K. Hall and W. M. Olsen (2001), 'Bridging the Gap Between Developmental Systems Theory and Evolutionary Developmental Biology', *BioEssays* 23: 10, pp. 954–62.

Schmidgen, H. (2015), 'Cerebral Drawings Between Art and Science: On Gilles Deleuze's Philosophy of Concepts', *Theory, Culture & Society* 32: 7–8, pp. 123–49.

Shea, N. (2011), 'Developmental Systems Theory Formulated as a Claim about Inherited Representations', *Philosophy of Science* 78, pp. 60–82.

Stengers, I. (2005), 'Deleuze and Guattari's Last Enigmatic Message', *Angelaki* 10: 2, pp. 151–67.

Sterelny, K. (2001), 'Niche Construction, Developmental Systems, and the Extended Replicator', *Cycles of Contingency: Developmental Systems and Evolution*, ed. S. Oyama, P. E. Griffiths and R. D. Gray, Cambridge, MA: MIT Press, pp. 333–49.

Sterelny, K. and P. Kitcher (1988), 'The Return of the Gene', *Journal of Philosophy* 85: 7, pp. 339–61.

Sterelny, K., K. C. Smith and M. Dickison (1996), 'The Extended Replicator', *Biology and Philosophy* 11, pp. 377–403.

Weber, B. H. and D. J. Depew (2001), 'Developmental Systems, Darwinian Evolution, and the Unity of Science', *Cycles of Contingency: Developmental Systems and Evolution*, ed. S. Oyama, P. E. Griffiths and R. D. Gray, Cambridge, MA: MIT Press, pp. 239–53.

Wimsatt, W. C. (2007), *Re-engineering Philosophy for Limited Beings: Piecewise Approximations to Reality*, Cambridge, MA: Harvard University Press.

Chapter 5

# Deterritorialisation and Creative Involution: A Note on Guattari and Deleuze

*Paul-Antoine Miquel*

*à Ivan Vukovic*

Deleuze and Guattari's collaborations contain a reflection on life which has both an ontological and a metaphysical sense. It is primarily a thought of the Earth, understood as a *ground without foundation* (*un fond sans fondement*), a thought for which the ground shifts *via* the machinic from the Earth to the Cosmic, so that the Earth appears as an *absolutely deterritorialised* ontological principle. It is thus not simply the human that is decentred by terrestrial gravity. The terrestrial itself oscillates continuously between Chaos and Cosmos. Through this perpetual movement of envelopment and expression, the real is divided into strata. Thus arises the problem of knowing what life consists of at the local level of the organic stratum, in a frame that is no longer absolute deterritorialisation but *relative*. Relative deterritorialisation means the diagrammatic lines of flight that make a stratum of the real irreducible to its assemblages, so that it thereby passes from one layer to another on the plane of consistency. From this second perspective, life concerns the articulation of the living being with the organic stratum. Examining the lines of flight proper to the organic stratum requires thinking the latter *beyond the model of the organism*. The Deleuzo-Guattarian perspective thus enters into debate with Lovelock and Margulis, who tend, on the contrary, to present the biosphere as a sort of super-organism.

On Deleuze and Guattari's model, that 'powerful germinal life' is not a simple principle of general metaphysics; it also has a heuristic and even epistemic value, which permits us to think the living being on the basis of life, as opposed to conceiving of life on the model of the living being. This life thus falls in line with a philosophical tradition more vitalist than organicist, which goes back to Boutroux and Bergson. But it has at the same time an undeniable scientific fecundity. That is at least what I will try to demonstrate.

In the first part of this chapter I will explain what deterritorialisation consists of, and why Guattari and Deleuze associate the image-concept of the Earth with it. The Earth is not simply the planet earth. It is an ontological principle, according to which there is no substantiality to the real. The real is always a ground without foundation or the correlate of itself, in its double dimension as plurality and genesis. The second part will show that Deleuze and Guattari, in contrast to what uncharitable readers have claimed, do not turn their backs on science. Science is not for them an artificial discipline. Rather, it touches on one ontological dimension: what they call the functional one. There is indeed a scientific plane of reference that must be placed in relation to the philosophical plane of consistency. For this reason, in the third and final part of the chapter, I will show that Deleuze and Guattari's model of 'non-organic life' and 'creative involution', far from breaking with scientific activity, can actually help us to better understand the fundamental changes to our notion of evolution wrought by symbiosis and horizontal gene transfer.

## The Earth, the Glacial, the Deterritorialised

What does it mean to say that Deleuze's thought is primarily a thought of the Earth, a *geology*? In the Western philosophical tradition, the Earth is opposed to the Heavens. The 'Sons of the Earth' evoked by Plato in *The Sophist* are Parmenides' adversaries. They want a philosophy without Paradise, without the Heaven of the Forms. They wish to think the world without *Hinterwelt*, without the world behind this one. But they are reduced ultimately to substantialising what they touch, what they observe and perceive. The real as such is reduced to the physical world. The 'Friends of the Forms', in contrast, conceive of what can be touched and seen in relation to a Heaven of Forms, the invisible site from which all sensible things derive their foundation and intelligibility. Even if in *The Sophist* Plato transforms the Forms into 'great kinds' and gives movement to the intelligible, he never goes back on the dichotomy of the sensible and the intelligible, or on the hypothesis that the former finds its foundation in the latter.

The Deleuzo-Guattarian way of thinking about the Earth is, on the contrary, a thought of *de*substantialisation. The Earth is a ground without foundation (*un fond sans fondement*), since it is impossible to conceive the Earth as ontological principle otherwise than through the event by which it is deterritorialised and reterritorialised. What comes first thus is not the Earth, nor *a fortiori* its human occupant, but this uncaused event that has, in a sense, always already taken place and that opens the way to an ontology of 'planes [*plans*]' and not of 'substance' (Deleuze and Guattari 1994: 155–6). At the heart of this ontology is the order-word 'PLURALISM = MONISM' (Deleuze and Guattari 1987: 20). To give this ontology a meaning or direction requires a metaphysics that advocates the univocity of Being or of the One:

> It is a fixed plane, upon which things are distinguished from one another only by speed and slowness. A plane of immanence or univocity opposed to analogy. The One is said with a single meaning of all the multiple. Being expresses in a single meaning all that differs. What we are talking about is not the unity of substance but the infinity of the modifications that are part of one another on this unique plane of life. (255)[1]

This passage indicates that the 'plane of immanence' is at the same time a 'plane of life'. Life must therefore be placed at the heart of the real – the life of the real. And starting to think life requires beginning with the Earth.

The Earth is the search for a foundation, but as such, it is already a figure of decentring. The Earth moves, it quakes. It is endlessly traversed by unmoving movements. The keys to these movements are to be found between *content* and *expression*, between *development* and *envelopment*. One such key is the word 'rhizome'. Paradoxically, to grasp the movement of the earth requires refraining from genealogy. Nomadology is not a monadology and geology is not a genealogy. The rhizome is an anti-memory, 'the opposite of a history' (23). It is possible to say, without distortion, that Deleuzian ontology demands thinking the Earth as a rhizome. Why this singular and anti-hereditary requirement? The rhizome is a multiplicity, but a smooth multiplicity from which the one is always subtracted (21). In order to understand the meaning/direction [*sens*] of this multiplicity, it's not enough to count its dimensions. If we start from the *n* dimensions of the rhizome in order to grasp its structure, we will never be successful. That's because the rhizome is not an ordinary physical object; it has a special topology. A rhizome is always *less* than the ensemble of its dimensions. We must not proceed by means of the association of elements to understand the structure of the rhizome but rather by transversality.

In mathematics, for example, it is not possible to treat the straight line (or even a finite segment of it) as a countable collection of points, since in so doing one could not account for the continuity of the line. Simply associating points among themselves never transforms them into a line. It requires Cantor's diagonal argument to achieve it.[2] It is possible, of course, to distinguish the points on the line, but it would also be possible to treat each point as a collection of spheres all converging on the same centre. Moreover, in talking about a straight line we are already far from the coast of Brittany. Could you treat the coast of Brittany as a straight line? No, as it happens. It certainly has a dimensionality, but only calculable if one accepts that it is not correct to say it's a simple straight light, or a plane, precisely because it is always *less* than a simple addition of dimensions: 1, 1+1, 1+1+1, . . . The key to coming to grips with the structure of a rhizome is not addition but rather subtraction: not $n + 1$ but $n - 1$. When it comes to the rhizome, the word 'structure' itself starts to pose a problem. The same goes for the word 'system'. It is less than a structure, less than a system. Why? For a simple reason: *what is structural to a rhizome is its incompleteness.* To say then that the Earth is a rhizome is to say that it is incomplete in principle. It forms a whole that is always already detotalised, always already absent from itself. My hypothesis is then that, for Deleuze and Guattari, *the Earth is a figure of decentring, not just in relation to the human, but in relation to itself.* This is the sense in which it is absolute deterritorialisation (56, 509). The Earth is always out of sync with itself, neither before, nor after, but *between* the elements of which it consists: 'the earth, the glacial, is the deterritorialised par excellence: that is why it belongs to the Cosmos, and presents itself as the material through which human beings tap cosmic forces' (509, translation modified). The Earth is always inhabited by a Chaos, so that it is not a ground [*fond*] except in the sense of being 'the deterritorialised', a correlate of deterritorialisation, and not in the sense of a foundation [*fondement*] (509). It is a ground without being essence or substance. For this reason, although inhabited by a Chaos, the Earth opens up to a Cosmos. In being always already out of sync with, or late for, itself, to the extent that it is inhabited by a Chaos, the Earth-rhizome is also always early. The more it is late, the more it's seized in a becoming-itself. In this way it becomes cosmic. The Earth, as ground, is a part of the Cosmos. It finds its foundation in a Cosmos that is not itself. But, inversely, the Cosmos is not a supernatural *Hinterwelt*, a Heaven. It is not vertical and hierarchical but horizontal, diagrammatic, machinic.[3] It falls under the jurisdiction of the event. The Cosmos is nothing more than that intensity, that movement at infinite speed, by which we, without being moved, realise that the Earth is not at the centre of the universe. It is nothing but the mark of the Earth's absolute deterritorialisation, at once with respect to us humans and also with respect to itself.

Thus it is that geology finds its foundation in cosmology, even though cosmology depends essentially on geology. *The Chaos/Earth/Cosmos complex is the event of itself*: not time but the interval or 'between-time', '*l'entre-temps*' (Deleuze and Guattari 1994: 158, translation modified).

The event is the mark that no spatiotemporal scale is eternal. The interval is, then, that of becoming, and the philosophy of the event, an ontology of becoming.

Consider a person running in place on a treadmill; the more the ground recedes before him, the earlier he must be to remain where he is. Here is a little image for coming to grips with the concept of '*l'entre-temps*'. The ontology of becoming is linked to the couple actual-virtual, and consistency is a sort of articulation of the actual and the virtual. How does the rhizome behave in order to be early for itself? It does so by being stratified. What gives it consistency are its stratifications, which is to say the movements through which it reterritorialises itself. In reterritorialising itself it is always ahead of itself; in the direction where there is nothing, it reconstructs, retotalises itself. This is the creative dimension proper to the rhizome. It invents itself each time in new directions in order to remain in place. It is thus that the One, in being *less* than all it *could be* is at the same time *always more* than what it is:

> It [the rhizome] constitutes linear multiplicities with *n* dimensions having neither subject nor object, which can be laid out on a plane of consistency, and from which the One is always subtracted (*n* – 1). When a multiplicity of this kind changes dimension, it necessarily changes in nature as well, undergoes a metamorphosis. (Deleuze and Guattari 1987: 21)

The One is at once always more and always less than itself. The One is not a principle but a correlate of the event of detotalisation and retotalisation, of development and envelopment, of content and expression.

To the extent that it is structurally *virtual*, deprived of itself, the One is multiplicity, transversality. One must not start from the One to understand the multiple, because the One is on the contrary the correlate of multiplicity. The concept of multiplicity thus becomes central for Deleuze and Guattari, even more than in Boutroux or Bergson, even if Deleuze avows the influence on his thought of *L'essai sur les données immédiates de la conscience*. But the One is simultaneously *active*, *creative* and *constructive*, since it is also the correlate of its genesis. It is always more than it is, because by this supplement there is always a becoming of its being that is irreducible to the being of its becoming. Here then are the two categories around which Deleuzian ontology is structured: transversality and becoming. The One is less than itself, it is the correlate of multiplicity; the One is more than itself, it is the correlate of its genesis. For this double reason, it never finds in its ground its foundation, and for this double reason the foundation of the One is not in its ground. That explains why, in order to come to terms with this ontology as a metaphysics, we must write the equation: 'pluralism = monism'.

## From Metaphysics to Science

In the philosophy of science, referring to Deleuze is usually disparaged, particularly in the reductionist contexts that dominate the field. It suffices only to recall the sadly famous Sokal affair that erupted in 1996. When Deleuze asserts with Guattari that the plane of immanence gives consistency to chaos, what does he mean by 'chaos' and by its 'infinite movements that turn back on themselves in incessant exchange' (1994: 42)? He writes for example that chaos 'undoes every consistency'. But any specialist in dynamic systems knows that deterministic chaos actually tends toward an attractor, the form of which it is sometimes possible to study – as in the case of the Hénon attractor (1976), an example of a 'strange attractor'. Sensitivity to initial conditions is certainly not the only criterion that makes it possible to define chaos. There is also that incredible property of topological transitivity, by which the chaotic system becomes an entirely new phenomenon that cannot be divided into two or three components (Devaney 1989). Moreover, there's the fact that the more a system tends toward chaos, the denser its periodic points become; in other words, the system has an infinite periodicity.

Deleuze and Guattari also say, not just of chaos, but also of the unlimited plane, that it is always 'fractal' (1994: 36), wishing to evoke no doubt that it is always less than the sum of its dimensions. But the dimensionality of a fractal is the object of mathematical calculations, which make it possible sometimes to characterise the structure of what physicists call a strange attractor. Here chaos doesn't escape scientific analysis. The fold, chaos, fractals: shall we conclude that Deleuze and Guattari are having a laugh at their readers' expense?

On the contrary, we must make an effort of reading: here we need to deal with philosophical frameworks and not scientific concepts. Deleuze and Guattari are constantly signalling as much to their readers. That said, one should not think these frameworks are metaphors. Deleuze and Guattari detest analogy and metaphor. The philosophical frameworks they call 'concepts' are 'vagabond, nondiscursive, moving about on a plane of immanence' (143). They are 'witch's flight[s]' (41). Joined to these frameworks are 'conceptual personae' which are not simple 'partial observers' like Laplace's or Maxwell's Demons (128–9). Lastly, philosophical schemes like the Cartesian '*cogito*', Spinozist 'immanence', Kantian '*Denken*' or indeed Bergsonian '*durée*' appear suddenly like singularities, 'black holes', in the midst of the plane of immanence, and lack any reference. When it comes to chaos, it is only a question of the 'virtual' dimension of the event, or, put more plainly, the virtual aspect of the real. It is this dimension that scientific activity diminishes as far as possible. The task of science is to retain nothing of the virtual but its actualisation, and hence it is fundamentally reductionist. Such reductionism, however, does not mean that science is arbitrary, artificial or just a matter of convention. Rather it stabilises us, lashes us to the Earth, but never manages to trace a witch's flight. This critical point, derived from Nietzsche, Spinoza, or again from Bergson, reveals something fundamental about this manner of practising philosophy. It does not deny the contribution of science to ontology. On the contrary. But it does aim to show that scientific reflection is not at the foundation of philosophical questioning. It is rather because the ground recedes continuously under the feet of the wise person that diagrams and abstract machines, refrains and witch's flights, are required, since these are the ways to move toward the Cosmos. This point is fundamental for what I will discuss below. My thesis is that there are lines of flight in the transition from the physical to the biological stratum, which justify making recourse to philosophical questioning and its conceptual schemes. These are the lines of flight that mark the expressive dimension of the real – not space or time but interspace and *entre-temps*; not form or matter, but expression and envelopment, the reversibility of interior and exterior, of thought and being.

Before developing this idea, let's return to the ontology of science, because *science has its own plane*, the plane of reference. Rather than a 'relativity of truth', it implies 'a truth of the relative' (130). Don't let the cookie-cutter formula obscure the point. If science has its own plane, then it is because there is already something of the absolute in science, a relative absolute, de-virtualised, but an absolute nonetheless. Why is it this way? *Because, as Deleuze and Guattari grasp perfectly, science de-subjectivises.* Science's blind spot is surely not just consciousness, but rather 'partial observers'. There is no science without a system-level perspective, which distinguishes itself from the point of view of any particular observer, which detaches itself from subjectivity. It is thus on account of the process of objectivation proper to all theoretical effort involved in scientific activity that the scientific absolute is relativised. There are no conceptual personages in science, but there are Demons, which attest to the way in which it is possible to characterise the states of the system in relation to its structure alone and no longer to an observing subject. These Demons open the door to the Cosmic, since they oblige human thought to get outside its own province, adding to the observer's perspective that of the system, which makes experimental science theoretical rather than simply empirical.

But in scientific practice, the door is closed. A theory makes reference to a system that it presents as closed on itself. *Science is only interested in assemblages and territories*, which is why it assigns importance exclusively to propositions, functions and states of affairs. If it's necessary to find diagrams, lines of flight, even in scientific practice, they will appear as blind spots, not only as a cross-section of chaos, but more profoundly as an effort to slow it down, to actualise the virtual, to spatialise becoming. It is the task of science to proceed thus – a noble task, but one must not confuse it with that of philosophy.

## The Organic Stratum: A Form of Relative Deterritorialisation

If the One is element of itself, if it is ground without foundation, if it is simultaneously enveloped and developed, then it is never completely folded or completely expressed. In every stratum and every territory, therefore, there are lines of relative deterritorialisation. Here I will focus on one among them, the organic stratum, in order to study it in more detail and show the specific sense in which Deleuze and Guattari give life to the organic stratum, intermediate between the plane of organisation and the plane of consistency. In the course of this analysis I will bring Deleuze and Guattari's thought into resonance with that of Lovelock and Margulis. Thus I aim to make evident the originality and contemporary relevance, not just philosophical but also epistemic, of their thought.

In Lovelock's work, understanding the Gaia Hypothesis and the ways in which the biosphere and the geosphere are woven together requires finding the lines of flight in the physiochemical stratum – hence the resonance with Deleuze and Guattari. Let's examine the point more closely. Lovelock is no reductionist, but neither is he an emergentist in the traditional Anglo-Saxon sense. It's not the autonomy of the biological domain that interests him. To grasp the interpenetration of biosphere and geosphere it's not necessary to begin with the biological in order to explain it on its own basis, but neither is it a matter of reducing the biological to physics. Rather one must seize on the lines of flight in traditional physical theories.

As a scientific hypothesis, Gaia is born from a preliminary 'recognition' of the presence of life on earth. Life is not observed but recognised, in the way that one surmises immediately that a sand castle on the beach is the work of a human being, whether a child or a parent. Such recognition rests neither on facts nor on hypotheses; it rests on signs of which one must interpret the sense. For example, take the simultaneous presence of large quantities of oxygen and of methane that react chemically together and rapidly disappear. This flagrant dissymmetry, a disparity in relation to thermodynamic equilibrium, is not sufficiently accounted for by means of the distinction between closed and open systems. The physics of phase transitions and dissipative structures tells us that a system that is open in the weak sense (energy exchanges) or in the strong sense (matter and energy exchanges) is capable of producing negative entropy and organising itself through interaction with its outside. But there are phase transitions on Venus and Mars, as well as processes of crystallisation and convection. And yet those planets do not support the simultaneous presence of methane, ammonia and oxygen. Everything is drowned in carbon dioxide. Here is what Lovelock calls *an anomalous* disparity (see the remarkable analysis in Dutreuil 2017). He thus unknowingly uses a vocabulary close to that of Deleuze and Guattari, when they evoke 'the anomalous' as a 'borderline' in relation to 'the pack or the multiplicity it borders' (1987: 245). The simultaneous presence of methane and oxygen is really a kind of thermodynamical *Moby Dick*. It makes no sense in the context of the banal physics of phase transitions. It is like a strange vanishing point, which is nevertheless immediately recognisable if we remember that the cycle of fermentation entails the production of methane and photosynthesis the production of oxygen. Everything becomes clear here once we perceive that the geophysical constants of the earth *are surrounded, inflected and regulated by life*. It's this anomalous disparity that will give Dr Eeger the chance to speak – a fictional personage, supposedly the author of the destruction of Gaia through the creation of an adulterous symbiotic alliance between a blue-green alga and a bacterium. Through this fictional story we will discover what's at the heart of the hypothesis: the biosphere regulates the physiochemical conditions of existence, through positive and negative feedback, like a kind of enormous thermostat. Yet it should be borne in mind that this thermostat is not the biosphere. It is the very theoretical assumption that the *biosphere comes from the geosphere, and that the geosphere is paradoxically an extension of the biosphere*: 'our contrasting view', Lovelock writes, 'required an atmosphere which was a dynamic extension of the biosphere itself' (1979: 7). Thus, Gaia is neither a physical entity, nor a biological one. As well noted by Latour (2015), she is not a sphere at all, because she is more and less than the biosphere. Gaia is simultaneously enveloped in the geosphere and developed in the biosphere as a ground without foundations.

In Deleuze and Guattari, biology is at the border of physiochemical strata. As in Lovelock, the biological is not thought in relation to itself. It is a kind of growth on the plane of consistency. It has no independence. It demands to be thought in a general cosmological frame, which the authors call 'double articulation' or again the 'double bind': B-A, BA (Deleuze and Guattari 1987: 40). And it is another conceptual personage, Professor Challenger, who tells the tale that resonates with that of Dr Eeger. What does the 'double bind' consist of? *The presence of a principle of transversality* in all strata (cf. above note 3). On account of this principle, nothing is ever first – not expression, nor the content that expression implies (Montebello 2008), precisely because being and appearance, the developed and the enveloped, are reversible.

Yet I want to note straight away that this principle is susceptible of having not only a general ontological meaning, but also an epistemic one and a local fecundity when it has to do with the scientific analysis of physical systems or biological complexes. It is at work as soon as there are lines of flight, of relative deterritorialisation, in the explication of natural systems. What does the double bind state? That B-A is not BA, but that nevertheless these two levels of reality are in communication (Deleuze and Guattari 1987: 40). One must think their communication from the point of view of B-A, that is, from the 'molecular' point of view, and also from the 'molar' point of view of BA. In more epistemological terms, one ought not to be content with either a bottom-up or a top-down mode of explanation. Everywhere we have to deploy the 'double bind' and everywhere it appears it is always a circle of which the centre is everywhere and the circumference nowhere. It is not made of points, it is between points. I insist on this: Deleuze and Guattari's hypothesis is not that we will find *everywhere* the molecular and the molar. That would only be a banal utterance, which would not at all explain the manifest existence in physical systems of forms of varying complexity. It consists rather of thinking that all forms of complexity are linked to a principle of communication between inside and outside, content and expression, implication and explication. It is thus that each form locally discloses the consistency of the real.

Deleuze and Guattari have a limited knowledge of physics. They only invoke briefly Simondon or Michel Serres. And yet it is possible to take up their distinction in more depth. Let's take just two examples from the field of thermodynamics. In statistical mechanics there is already a double bind, already something that is not simply of the order of content but also of coding. Take the temperature of a gas. I cannot assign a temperature value to each atomic element that composes it. That would be absurd. Temperature is a statistical value one can only assign to a population of atoms. Even, then, in statistical mechanics, there is already a double bind, a B-A, BA; temperature is nothing other than an emergent property of a population of atoms, which is retrospectively defined as a quantity of agitation proper to each of them. Temperature at the level of a single atom is a notion that makes no sense.

Indeed, we always see in physics that the molecular signification of the double bind has nothing to do with the molar signification. Take now the example of a phase transition that, in interaction with the exterior context, manifests, for certain values of the control parameters of a given physical system, a global property like the magnetisation of iron. This property is no longer seized upon from the outside by the theorist, who cannot observe the atoms of a gas one by one. It emerges as a *haecceity, a singularity* in interaction with the exterior milieu, and it regulates, in return, the interactions between each element composing the physical system. There is indeed a BA, and yet it depends on conditions that it is possible to articulate and even determine exactly at the level of the B-A. The double bind is here again: expression/content at the molar level as at the molecular one.

It is possible to develop this in a more technical way and to show how the principle of transversality assumes a strong form in the physics of phase transitions. The technical term to which it would be linked is the principle of the non-decoupling of scales [*le principe de non découplage des échelles*]. It is impossible for a thermodynamic system approaching its critical point to find a level of explanation where the interactions between its elements disappear. Whatever the level, interactions or, to speak like a physicist, correlations turn up. To understand what occurs, it is therefore necessary to *place oneself between the B-A and the BA*, to look for a manner of describing that system that makes it possible to understand what is occurring, not at one scale, but between scales. In the language of Deleuze and Guattari, it is necessary to find the lines of flight, or a diagonal point of view. Procedures exist, like those dealing with renormalisation groups, that make it possible in this way, not to describe a given physical system directly, but rather to describe the relation between the parameters defining the system at one scale and those defining it at another, and so on, moving toward a fixed point for these descriptions, thanks to recursive equations (Wilson 1983). It's like studying the relation between distinct objects in several photographs whose scales are changing, in order to capture an invariant property that will always be present when the thermodynamic system passes from one state to another (Castiglione et al. 2008; Longo and Montévil 2014).

The biological stratum is not independent of the physical one; there is a continuous process leading from one to the other. And yet there is at the same time rupture, a difference of degree transformed into a difference of kind. I want to suggest at the moment that *this is what produces lines of flight in the organic stratum itself*. That seems to me to be the originality of Deleuze and Guattari in relation to Lovelock. It is no longer a matter of showing what classical physical theories lack when it comes to understanding the organic stratum. Rather, it's a matter of showing that there are indeed points of local deterritorialisation in the organic stratum itself. *That requires explaining why, at the local level, and in the organic stratum, germinal life exceeds and goes beyond biological organisation.*

Specific to the organic stratum is the idea of an encoding that unfolds on an independent 'line of expression' and frees itself from spatiotemporal dimensions (Deleuze and Guattari 1987: 59) in order to become *functional*. Thus we move from simple diagram to program, or again from physics to biology. There is of course always a difference of kind between the two, since physical space is doubled here by a space of biological functions. That occurs as much at the molar as at the molecular level, and Deleuze and Guattari are fully aware of it, even if their texts are not always entirely clear about it. At the molecular level, DNA is already functional. It is replicated, transcribed and translated into amino acids, by means of messenger, transfer and ribosomal RNA. It is evidently the same at the level of the living cell. To take just one example, a bacterial cell divides in the same way that DNA is replicated. But we know all the same that it feeds itself and repairs itself. It fulfils certain functions. There's already a division of physiological labour within the cell. There is molar encoding. Still, here it's a matter of an oversimplified division of labour. The 'molar' and the 'molecular' can also be found between biochemical metabolism and biological macromolecules, as Guattari and Deleuze correctly point out (59).

The concept of a 'program', however, encourages one to imagine a hierarchical relationship here: an independent line of expression now governs the processes of encoding and decoding, as if every living cell was already inscribed implicitly in the DNA of its chromosomes. Is it not therefore necessary to say that biology is always already reterritorialised on the organic? Doesn't the genome thus specify the living organism, since it contains the latter's program or blueprint? A certain way of thinking dominates the era of Deleuze and Guattari, which can be found in Monod and Jacob. In *The Growth of Biological Thought*, Ernst Mayr (1982) even makes of the genetic program the equivalent of an Aristotelian formal cause. If such was the case, however, our strictly organismal understanding of living things would rest on the concept of a hierarchical or teleological plan, a plane of transcendence (Deleuze and Guattari 1987: 265). And yet what matters for Deleuze and Guattari is *to think the organism as correlate of the biological stratum* and not *vice versa*. Organisation is not at the centre of their thinking, and the holistic approach is not the answer in biology. At the biological level, as at the physical, there are no strata without 'epistrata' and 'parastrata'.

Let's take two classic examples. First, plants like bamboo and fern show the point at which the traditional notion of organism actually poses a problem for biology. To begin with, a fern is not a mammal. Besides sexual reproduction, a fern can reproduce by vegetative multiplication, or cloning, thanks to its rhizome. If we define an organism in advance by its physiological unity, where is the organism in this case? At the level of each fern or of the colony? To clarify our ideas, let's add that among ferns sexual reproduction forms a cycle. It starts with the sporophyte organism, which produces spores. Once fallen on the ground, they sprout into a second gametophyte organism, the prothallium, which produces in turn male and female gametes that fertilise one another. The product of this fertilisation will issue in a new fern. Where is the organism that reproduces itself? Is it sporophyte or gametophyte? This first example clearly calls into question the physiological unity of the biological organisation.

Take the eukaryotic cell as a second example. If what it is to be an organism is defined no longer as a physiological unity, but on the basis of the program contained in its genome, we fall immediately on a second structural problem: mitochondria in the eukaryotic cell also contain genes. Genes are not restricted to the cell nucleus. Today the problem is explained by a conjectural endosymbiosis between an archaeon and a eubacterium. The archaeon would have phagocytised an aerobic bacterium, then a green algae-type bacterium capable of producing oxygen by photosynthesis. In this way, the hypothesis goes, the mitochondria and chloroplasts of modern eukaryotic cells were born. But where then is the organism, and where are its programs, in the nucleus of the cell or in its cytoplasm instead? In a sense, the cytoplasm is part of the environment of the eukaryotic cell nucleus, but in another sense, yeast and trypanosoma are unicellular organisms in their own right. How is it possible to decide the question? Again, we can do no other than to suppose, for the moment, that the independence of an expressive genetic line is sufficient to define an organism – something that is wholly false! Deleuze and Guattari are victims of the perspective of their time. The idea that there is a process of encoding and decoding at work in living macromolecules means nothing more than that genes are an independent expressive line. There is no genetic stratum without epigenetic, proteinous, cellular and material parastrata, or without ecological epistrata. That's because, as Lewontin (2001) writes, the whole organism is defined by the fact that it is the correlate of a co-construction with its environment. Here's the reason why the whole organism contains the rhizome and symbiosis as lines of flight. To be sure there are entities that are more like colonies, like ferns and corals, and those that are more unitary, like mammals, but there is always something of the colony folded within the unitary. Bacterial endosymbionts in mammals are crucial to their fundamental biological functions: digestion, but also development, immune system function, and even thought. And even if we abstract a mammal from its bacteria, it is certainly not simply at the level of the genome, the genomic stratum, that we can define its organisation. *An organisation is always a correlate of its transversality. Transversality is not the result of the organisation.*

On this basis where is the line of flight? Let's make an effort to get hold of it: *one cannot think the living without life*. The life of living things is primary. A living thing, an organism, is only a *modus vivendi* – but a vortex, a black hole, a haecceity, stitched together with all the others on the same stratum and in the midst of the same plane of consistency. The line of flight is life itself. One only comes to grips with it at the level of the biosphere, not the organism: *the biological stratum* par excellence *is the biosphere of which organisms are only correlates. Life has an expressivity that overflows biological organisation.*

## Creative Involution

It is important to focus on the originality of Deleuze and Guattari's thought. It tells us something about life that Lovelock isn't able to articulate. Life is not simply biological organisation, and the biosphere cannot be compared to a sort of enormous organism.

One can indeed affirm with Lewontin that an organism is simultaneously cause and effect of environmental modifications, and that likewise the environment is cause and effect of the modifications of organisms (2001: 66). An organism does not passively submit to its milieu without at the same time constructing it. And inversely the development of an organism is always something in which the environment participates. The traditional division of labour between developmental and evolutionary biology must therefore be rethought. The decoupling of ontogeny and phylogeny is not as clear and distinct as it was in the era of triumphant neo-Darwinism. Nevertheless, there is always a decoupling between ontogeny and phylogeny. That is characteristic of life. We can never say of an organism that it has evolved. An organism is robust in certain respects but in others it is plastic. It has a certain adaptability, and so in a sense an evolvability. *But it is the biosphere that evolves, not the organism.* With organisation stability dominates, not plasticity. On this point we can refer to the recent work of Montévil and Mossio (2015). In contrast, what dominates the biosphere is plasticity. Here is the central point that Lovelock's model prevents us from grasping. Why is it thus, and through what philosophical framework is it possible to come to grips with the plasticity of the biosphere? The classical, traditional framework is that of evolution.

For starters, we insist on the Darwinian innovation which Deleuze and Guattari mostly fail to mention in *A Thousand Plateaus*. The traditional ways of conceiving of classifications, since Richard Owen, emphasised structural homologies, rather than functional analogies. Thus dolphins, bats and human beings are mammals, while bats and birds don't belong to the same family of species. Georges Cuvier challenged this way of thinking, and his controversy with Geoffroy Saint-Hilaire is well known. For Cuvier it was physiology and not anatomy that held the key to a good classification. He thus emphasised the function rather than the organ. A carnivore has the jawbone, neck muscles and teeth of a carnivore. It's the function that matters. Darwin completely reverses these two perspectives in *The Origin of Species* (1859). How do you know what homing, Jacobin and Pouter pigeons have in common? Common traits such as the dark marks on the plumage of all these domestic breeds encouraged him to believe that they all derive from the same ancestor: *Columba livia*. All Darwin's cleverness and acumen is devoted to transforming *genus* from the abstract class it had been for traditional classification into a matter of *common ancestry*, to the extent that logic becomes a ground that *now finds its foundation in chrono-logic*. Time and becoming come first, not class. This is Darwin's great lesson. In fact, I maintain that Darwin's most 'dangerous idea' is not, as Dennett argued (1995), the principle of natural selection, but rather *the principle of descent with modifications*. Only by this principle can we understand that the action of selection isn't simply eliminative, in contrast to what the first population geneticists would have had us believe in the twentieth century. On the contrary, *selection acts as a diversification of species*. What it maximises is not 'fitness', but the principle of 'divergence of character'.

Why is that so? Because of the principle of descent with modifications, which introduces hereditary variation alongside the action of selection. *This is a non-conservative principle*,[4] which attests to the fact that what comes first is not class or genus but genesis or becoming. And it is this principle that makes us move from a fixist perspective to an authentically evolutionary one. We can use the principle to return to our central preoccupation: the biosphere can be represented as a tree with continuously ramifying branches (Darwin 1859: 116–17). In reality, as it happens, when we examine the Darwinian schema more closely, it resembles not one tree but many distinct trees, such that no function can bring together these common sources. On this basis, the stochastic dimension is embedded in life *from the origin*. A continuous function cannot represent the tree of evolution, even if the applications of a given function to itself diverge more and more over the course of time.

Gould's punctualist schema notably retains the same general structure of explanation. The tree reappears. Now it is necessary only to take into account how the major branches diverge in a discontinuous way in the Precambrian, the moment of the sudden explosion of distinct life forms (Gould 1989: 46–7). Natural selection acts only in the second place, through decimation. It eliminates certain lines and causes little branches to appear on the bigger ones. But there is no further explosion, as in the Precambrian. By natural selection vertebrates, and then mammals, derive ultimately from the eel-like Pikaia.

But is it necessary to think of evolution as a tree? This is the central question Deleuze and Guattari ask. We are now in a position to measure at what point those who derided them ought to swallow their insults.[5] Deleuze and Guattari challenge us, in contrast, to conceive of evolution as a 'creative involution'. For that it is not sufficient to say that the biosphere evolves without also taking care to grant *that evolution itself has a rhizomatic structure*. In evolution, the biosphere does not appear simply as the correlate of its genesis. It appears at the same time as the correlate of its heterogeneity. *It is structurally incomplete.*

Evolution is an involution. Why? Because there is a principle of heterogeneity at work in the biosphere: the whole eludes its totalisation. It is opened up from the inside, and evades its own closure. It constitutes itself as such through collapsing in [*effondrement*] on itself. *It is opened up from the inside, because the whole is less than the multiplicity that constitutes it. This is the content of the principle of transversality.* Evolution, as involution, is thus only the most visible, explicit manifestation at the level of the organic stratum's relative deterritorialisation of the cosmic principle of absolute deterritorialisation. The biosphere, essential dimension of the Earth, is the correlate of the principle of transversality. Through it the Earth appears thus necessarily as a ground without foundation. In this case, however, there's another side to the coin. If the biosphere is always less that what it could be, by application of the principle of transversality, *it is at the same time more than it is. The deterritorialised is continually reterritorialised.* Its constructive dimension arises out of the fact it collapses in on itself [*son effondrement*], such that involution can also be creative. Life appears then as the double correlate of both its transversality and its genesis. The fact that it is continually detotalised creates the conditions for the movement through which it is retotalised: 'it amounts to the same thing to say that each multiplicity is already composed of heterogeneous terms in symbiosis, and that a multiplicity is continually transforming itself into a string of other multiplicities' (Deleuze and Guattari 1987: 249). Principle of transversality, principle of transformation: this is a new chrono-topological vision of evolution.

In order to think evolution it is necessary to depart from the model of filiation and heredity, and to return to the model of 'alliance' and 'communication' (238). Three years after the gruesome Sokal Affair, W. Ford Doolittle will publish an article in *Science* in which the notion of a 'reticulated tree' first appears (1999), on which that of the 'ring of life' will closely follow (Rivera and Lake 2004). The latter notion makes it possible to imagine that the birth of eukaryotes should have been accomplished not by filiation but by the symbiotic fusion of living cells from the *achaeae* and others from among the *eubacteria*. The topology of the 'ring of life' presupposes the functioning of the principle of transversality, the properties of which will not cease to be studied, and which are notably linked to the importance of endosymbiosis and lateral gene transfer, made possible by different mechanisms like retroviral infection. Clearly, then, there exist avenues of deterritorialisation, transversal avenues that continuously nourish a certain kind of becoming proper to life on earth, a becoming, a genesis, through which it is not the same that returns but the other, not repetition, but creation.

Life has therefore a special meaning, in Deleuze and Guattari, and a general one. From the general perspective, the Earth, on the ontological level, must be thought of as a correlate of deterritorialisation. The Earth as decentring of the human, which does not find its own ground in itself, is itself decentred in the Cosmos, and this is what philosophy must take hold of: this dimension of decentring of the Earth in the cosmos. The Earth is never substance, since it is an intermediary between envelopment and development, content and expression, genesis and multiplicity. The Earth is never itself without being more or less than itself. This is the general sense of life as an ontological principle. I have attempted to show that this general sense resonates with its special sense on the organic stratum. I also wanted to show that, despite what certain readings of Deleuze and Guattari may have suggested, their special sense of life has a contemporary epistemic relevance that is not alien to the development of scientific activity, even if it is not situated on the same plane. To put it more directly, this conception of life is likely to have scientific productivity. It simply says that, in order to understand life, it is not enough to extend and enlarge one's attention to the physiochemical conditions that have been able to produce it, as Lovelock does. It is necessary to think the biological otherwise than on the organismic model. In the biosphere there are lines of flight, a specificity that obliges us to leave behind the organism as model, and to think the living being beginning from life itself once again, and not vice versa. The biosphere not only evolves, as Darwin said, but involves. It involves because it is necessary to understand it not simply on the model of hereditary filiation but on that of horizontal transfer – the alliance model. *The biosphere involves because it is at the same time open on itself*. That is the lesson we should learn from these two authors.[6]

## Notes

1. For a more precise examination of the connection between the planes of univocity, immanence and consistency throughout Deleuze's philosophy, see Montebello 2008 and Lapoujade 2014.
2. Suppose that the real numbers have the same power as the whole numbers. They must therefore be countable. We should be able to draw up a list (r1, r2, r3, . . ., rn, . . .) of all the real numbers between zero and one. Now take the diagonal that consists of the first decimal of the first of these numbers increased by 1 (modulo 10 in decimal notation), then the second, the third, always increased by one, and so on. We have now therefore constructed a new number that could not have been on the list, since it is not equal to any element of the list (r1, r2, r3, . . . , rn, . . .). It will be different from each element by at least the value of one of its decimals. *But at the same time this number is indeed between zero and one*. This is a very simple demonstration of the lack of bijective correspondence (isomorphism) between the set of whole numbers and the set of reals. There is an infinity of numbers in the first set, but they are countable, which is not the case with those in the second. No rule of correspondence will make it possible to project the set of real numbers onto that of natural numbers. The first are more 'numerous'.
3. The terms 'horizontal' and 'transversal' designate the rhizomatic dimension of the Earth, which is pure multiplicity. But the notion should not be taken in a literal sense. The word 'transversal' has a metaphysical sense. It means that the Earth as an ontological principle is less than the ensemble of its dimensions.

*For this first reason, the Earth is not identical to itself, or other than itself.* It is effectively defined by the fact that it is not possible to totalise it. Every form of unity (the One) is less than this de-totalisation. But as multiplicity irreducible to the one, the Earth is by this very fact *continual potential and change*. Becoming is thus nothing but the actualisation of this potential for change.

4. On this point see the illuminating analyses of Longo and Montévil (2014), as well as those of Soto et al. (2016) on the relation between descent with modifications and the proliferation of variants.
5. See the notably lucid treatment by the analytic philosopher and connoisseur of Deleuze, Pascal Engel (2001).
6. I would like to thank Mike Bennett for translating this chapter for inclusion in the collection and Jean-Marc Lévy-Leblond for reading over the finished manuscript.

## References

Castiglione, P., M. Falcione, A. Lesne and A. Vulpiani (2008), *Physique statistique: Chaos et approches multi-échelles*, Paris: Belin.

Darwin, C. (1859), *On the Origin of Species by Means of Natural Selection*, London: John Murray.

Deleuze, G. and F. Guattari (1987), *A Thousand Plateaus: Capitalism and Schizophrenia*, trans. B. Massumi, Minneapolis: University of Minnesota Press.

Deleuze, G. and F. Guattari (1994), *What is Philosophy?*, trans. H. Tomlinson and G. Burchell, New York: Columbia University Press.

Dennett, D. C. (1995), *Darwin's Dangerous Idea: Evolution and the Meanings of Life*, New York: Simon and Schuster.

Devaney, R. L. (1989), *An Introduction to Chaotic Dynamical Systems*, Boulder, CO: Westview Press.

Doolittle, W. F. (1999), 'Phylogenetic Classification and the Universal Tree', *Science* 284, pp. 2124–7.

Dutreuil, S. (2017), *Gaïa: hypothèse, programme de recherche pour le système terre, ou philosophie de la nature?* Doctoral Thesis, Institut d'Histoire et de Philosophie des Sciences et des Techniques, Paris.

Engel, P. (2001), 'L'affaire Sokal concerne-t-elle réellement les philosophes français?', *Philosopher en français*, ed. Jean-François Mattei, Paris: Presses Universitaires de la France, pp. 557–76.

Gould, S. J. (1989), *Wonderful Life: The Burgess Shale and the Nature of History*, New York: W. W. Norton.

Hénon, M. (1976), 'A Two-Dimensional Mapping with a Strange Attractor', *Communications in Mathematical Physics* 50, pp. 69–77.

Lapoujade, D. (2014), *Deleuze, les mouvements aberrants*, Paris: Minuit.

Latour, B. (2015), *Face à Gaia*, Paris: La découverte.

Lewontin R. (2001), 'Gene, Organism and Environment', *Cycles of Contingency: Developmental Systems and Evolution*, ed. Susan Oyama et al., Cambridge, MA: MIT Press, pp. 59–66.

Longo, G. and M. Montévil (2014), *Perspectives on Organisms: Biological Time, Symmetries and Singularities*, Berlin: Springer.

Lovelock, J. (1979), *Gaia: A New Look at Life on Earth*, New York: Oxford University Press.

Margulis L. and R. Fester (eds) (1991), *Symbiosis as a Source of Evolutionary Innovation: Speciation and Morphogenesis*, Cambridge, MA: MIT Press.

Mayr, E. (1982), *The Growth of Biological Thought: Diversity, Evolution, and Inheritance*, Cambridge, MA: Harvard University Press.

Montebello, P. (2008), *Deleuze: la passion de la pensée*, Paris: Vrin.

Montévil, M. and M. Mossio (2015), 'Biological Organisation as Closure of Constraints', *Journal of Theoretical Biology* 372, pp. 179–91.

Rivera, M. C. and J. A. Lake (2004), 'The Ring of Life Provides Evidence for a Genome Fusion Origin of Eukaryotes', *Nature* 431, pp. 152–5.

Soto, A., G. Longo, P.-A. Miquel, M. Montévil, M. Mossio, N. Perret, A. Pocheville and C. Sonnenschein (2016), 'Toward a Theory of Organisms: Three Founding Principles in Search of a Useful Integration', *Progress in Biophysics and Molecular Biology*, 122: 1, pp. 77–82.

Soto, A., G. Longo, M. Montévil and C. Sonnenschein (2016), 'The Biological Default State of Cell Proliferation with Variation and Motility, a Fundamental Principle for a Theory of Organisms', *Progress in Biophysics and Molecular Biology* 122: 1, pp. 16–23.

Wilson, K. G. (1983), 'The Renormalization Group and Critical Phenomena', *Review of Modern Physics* 55: 3, pp. 583–98.

Chapter 6

# Hydrosocial Becomings: Evolutionary Perspectives on Water Assemblages and Maya Kingship

*Johan Normark*

Although archaeology has been used in a metaphorical way (i.e. by Foucault 1969), few philosophers take actual archaeological research seriously.[1] Deleuze and Guattari are exceptions in this regard since they discuss the work of Gordon Childe and André Leroi-Gourhan (Deleuze and Guattari 1987: 60–4, 428–9). Few archaeologists have followed Deleuze and Guattari's treatment of these particular archaeologists (Normark 2015b). In archaeology Deleuze has primarily been used in studies of things and assemblages (Hamilakis 2013; Harris 2014; Lucas 2012; Normark 2006, 2010).

The development of the discipline of archaeology is closely intertwined with evolutionary theory (Johnson 2010; Shennan 2002). As one might expect from the paragraph above, most evolution-oriented archaeologists do not discuss Deleuze and Guattari's (1987) rhizomatic and machinic perspectives on evolution, which detach the impetus for change from a particular part of the organism (like the gene in neo-Darwinism). Evolution for Deleuze and Guattari is not restricted to the hereditary transmission of traits, reproduction, replication and selection. It involves a changing field of multiple assemblages of heterogeneous components (Ansell-Pearson 1999: 171).

Some of Deleuze's followers use concepts like flow and fluidity to describe the becoming of life and the world (DeLanda 1997). At least since Heraclitus, flowing water has been used as a metaphor for change and transformation (Neimanis 2012; Normark 2015a; Strang 2014). Organic life as we know it emerged in liquid water and it has been crucial in the evolution of organisms ever since. As assemblages of minerals, carbon-based molecules and water, organisms took a major evolutionary leap when they left a predominantly aquatic domain. Most terrestrial organisms still need freshwater for survival, and for this reason freshwater has affected the evolution of *Homo sapiens*. The spread of humans across the earth has likewise been driven by climate change and locations of freshwater. The search for scattered sources of freshwater has led to the development of longer limbs and a lighter body to improve terrestrial mobility (Finlayson 2014). One of the limitations for humans spreading beyond earth out into the solar system is the lack of easy access to sufficient volumes of liquid freshwater (Normark 2015a).

Water has affected the evolution of not just the human phenotype, but also of human settlements, construction materials, socio-political organisations, agriculture, ceramic production, transportation, painting, etc., all of which have left archaeological remains (Normark 2014). To paraphrase Ingold and Palsson (2013), these remains are examples of *hydrosocial becomings*.

In a general sense this essay aims to show how Deleuzo-Guattarian concepts of becoming and evolution provide other perspectives than the more common neo-Darwinian and orthogenetic evolutionary perspectives on culture. The specific aim of this essay is to show how water acts as a catalyst in the becoming of multi-scalar hydrosocial assemblages. The Prehispanic Maya culture provides empirical examples of how substantial changes occurred in relation to water. Droughts affecting water management became one of the prime movers in the infamous 'Maya collapse', when the Maya kingship ceased to exist one millennium ago. Hence, water and its various assemblages are useful for discussing the relevance of Deleuzian concepts in archaeology and anthropology.

## Social and Cultural Evolution in Archaeology

Before I discuss this collapse in more detail, a brief summary of the use of evolutionary ideas in archaeology and anthropology is in order. Although *social* and *cultural evolution* are concepts used by some archaeologists, they are flawed since both rely on the Nature-Culture dichotomy (Descola 2013; Ingold 2000). As is the case in the 'mother-discipline' of anthropology, archaeology has by tradition been based on the idea that a singular society or culture (Greek culture, Maya culture, etc.) is the result of a specific arrangement between the modernist terms Nature and Culture. Generally speaking, archaeology studies how these cultural patterns emerge, evolve and disappear. I shall not problematise the concept of culture too much in this text but, as I have argued elsewhere, it is often an arborescent concept, acting as a master signifier that orders entities in a hierarchical order (Culture → Maya culture → Maya pyramid, etc.) (Normark 2006).

Rather than discarding the concept of culture I shall follow *dual-inheritance* theorists Richerson and Boyd's argument that culture is 'information capable of affecting individuals' behavior that they acquire from other members of their species through teaching, imitation, and other forms of social transmission' (2005: 5). From an archaeological perspective, such information emerges in people's interaction with artifacts, architectural features, landscape modifications, etc. These entities act as scaffolds of an extended and enactive mind (Clark 2003; Malafouris 2013; Normark 2016b).

Historically, cultural evolution studies have focused on classifying social complexity rather than on information. In the late nineteenth century Herbert Spencer popularised the idea that societies evolve toward more complex organisations (Johnson 2010: 148–9). This early form of teleological evolutionism disappeared in favour of the idea of culture in the singular, where the diffusion of cultural traits became more important in archaeology (known as culture-history). However, anthropologist Albert Kroeber defined these singular cultures as super-organic entities (Descola 2013: 74). Organismic metaphors for cultural forms have been common in various social sciences (DeLanda 2006), and this is one reason why social and cultural evolutionary perspectives have remained common in macro-scale studies. Indeed, Lyman and O'Brien (2001) argue that American culture-history relies on Darwin's descent-with-modification perspective in which contemporary cultures have descended from previous ones.

The mid-twentieth century saw a return of evolutionary studies on social complexity with teleological implications. Anthropologist Leslie White's use of thermodynamics suggests that social complexity emerges through the extraction of greater quantities of energy (Johnson 2010: 150–1). This perspective became more popular in archaeology than in social anthropology and it favours an *orthogenetic* perspective on evolution where sequences of ideal forms follow each other through predefined stages rather than directly descending from previous forms (Lyman and O'Brien 2001; Richerson and Boyd 2005). Examples of such unilinear evolutionary stages include Elman Service's band, tribe, chiefdom and state. Morton Fried has proposed similar stages of complexity: egalitarian, ranked, stratified and state (Johnson 2010: 153). These versions of cultural evolution have been criticised by postmodern archaeologists for being too monolithic and teleological (Shanks and Tilley 1987). Despite such criticism, these perspectives have remained important in complex systems-based archaeological perspectives that maintain the evolutionary stages for classification of social complexity.

Other evolutionary perspectives are grounded in neo-Darwinism (Shennan 2002). In neo-Darwinism genes or genotypes have replaced organisms as the entities being inherited, selected and replicated (Ramirez-Goicoechea 2013: 59). Apart from genetic inheritance there is also epigenetic, behavioural and symbolic inheritance (Fuentes 2013: 51). Neo-Darwinian cultural evolution often replaces genotypes with cultural types as units of inheritance (and material cultural types in archaeology). Cultural types have the properties needed for selection, such as 'variation, competition, inheritance, and the accumulation of successive cultural modifications over time' (Lewens 2015: 14). Skills, knowledge and linguistic expressions are passed between generations through learning and imitation. These capacities can also be passed between people through cooperation and mutuality (Fuentes 2013: 57; Lewens 2015: 10).

One of the outspoken critics of neo-Darwinism is anthropologist Tim Ingold (2013). He is also an advocate for the relevance of Deleuzian and Bergsonian insights into evolution for anthropology. I agree with Ingold's argument that evolution is a process grounded in *ontogenesis*, which describes how the milieu affects the developmental process. According to Ingold, neo-Darwinism has largely ignored the ontogenesis of the phenotype. The 'final' or adult form has been conceptualised as already pre-existing in the genes before development. Ontogenesis has therefore been reduced to a transcription of a genetic code (Ingold 2013: 6). An important factor in ontogenesis is the *morphogenetic field* that explains embryological development (Beloussov 1997). This field consists of cells that develop specific structures in response to biochemical signals, which means they are not determined by a pre-existing essence. Species are not timeless essential categories; they are historically constituted entities which resemble each other because they have passed through similar immanent ontogenetic processes. Manuel DeLanda (2002: 10) expands these morphogenetic fields beyond embryology and biology into multi-scalar assemblages. The chapter's next section will attend to these multi-scalar assemblages and elaborate upon two complementary approaches inspired by Deleuze: DeLanda's assemblage theory and Levi Bryant's machine-oriented ontology.

## Populations and Multi-scalar Approaches

Cultural patterns emerge from the way people and other entities interact collectively, and if cultural evolution is to be cumulative then there must be inheritance of cultural traits on the population level (Richerson and Boyd 2005). An individual does not inherit or acquire all the knowledge and skills within a culture. Only in larger assemblages, or in multiple assemblages, can the accumulation of cultural knowledge and skills emerge and be maintained. Hence, from a cultural perspective, continuity of practices and knowledge is found in populations rather than in individuals.

According to Ingold, population perspectives are the opposite of relationist perspectives (including those of Deleuze and Bergson). In Ingold's own words, population thinking 'has always defined the neo-Darwinian project, according to which every individual is a discrete, bounded and enumerable entity, one of a population of such entities, and relating to other such entities along lines of contact that leave its internally specified nature unaffected' (2013: 13). However, not all population approaches treat individual entities as discrete units whose interiors are unaffected by external relations. DeLanda's (2006) assemblage theory differentiates between relations of *interiority* and *exteriority*. In organismic models of culture, relations of interiority exist to keep the culture alive, discrete and homogeneous. In contrast, relations of exteriority exist between differentiated entities within a population. Humans form relations of exteriority with tools, buildings and crops, and all evolve as assemblages.

Ontogenetically, human individuals are assemblages that emerge from the interaction between internal components and components external to the subject. The capacities of an assemblage create opportunities, resources, constraints and risks for its components (DeLanda 2006: 33–4). Different entities can be formed from similar units, and assemblages of one emergent order can be less complex than their own parts. A social assemblage is usually less complex than the entities it consists of. For example, a family is less complex than each of the family members' biochemistry.

Ingold further suggests that 'the species concept, employed as a biological taxon, is a product of population thinking' (2013: 18). In DeLanda's (2002, 2006) account it is typology that has primarily defined species, not population thinking. Typological species concepts go back to pre-Darwinian – i.e. Aristotelian, Linnaean and Cuvierian – classification systems where all species have been the same since creation due to their essential properties (de Queiroz 2005). In the typological approach, a transcendent and essential type stands for a whole population and individual differences are glossed over. Here *Homo sapiens* is a species that represents billions of individual entities.

Similar to Darwin and population thinkers, DeLanda (2002) views species as the result of immanent differentiation, not a realisation of transcendent types. To phrase it differently, the human individual is the *actualisation* of a *virtual body plan* which, contrary to Cuvier's typological body plan, consists only of differential relations between pre-individual singularities. It's not 'given' in advance, nor does it resemble its actual products. Although all human individuals have different appearances they share a common body plan consisting of distributed attractors which create various forms through bifurcation. Through such bifurcations an organism is ontogenetically developed (DeLanda 2006; Normark 2012).

DeLanda treats the concrete as actual and structure as virtual, but in a multi-scalar view what is virtual on one scale may be actual on another. For Bryant (2011), the virtual is part of each concrete entity. Hence, 'for DeLanda the virtual is *external* or behind and beneath [but] for Bryant it is *internal*' (Di Liberto 2015: 124, emphasis original). Bryant's (2011: 69) *virtual proper being* of an entity is its substantial dimension, a difference engine that makes it an entity without following a predefined schema. Since each entity is embedded in other entities, DeLanda's morphogenetic field is the virtual proper being of a larger entity (Di Liberto 2015: 125). Entities generate events or actualisations (*local manifestations* in Bryant's terminology). This occurs when entities form *exo-relations* (external relations) with other entities (Bryant 2011: 69). Thus, entities are both clusters of relations and individual units (Di Liberto 2015: 127).

In a later book Bryant uses the term *machine* 'to emphasize the manner in which entities dynamically operate on inputs producing outputs' (Bryant 2014: 6, emphasis removed). The input is affected by the *endo-relations* (internal relations) of the machine (Di Liberto 2015: 114–15). Whereas DeLanda (2006) defines assemblages from their properties and capacities, Bryant (2014) defines machines from their *powers* to act upon other machines. From here and onwards I shall use Bryant's definition rather than DeLanda's since it emphasises how a machine acts on other machines. Powers are virtual and therefore belong to the endo-relations of a machine. However, powers are manifested through exo-relations and always exceed their manifestation (Di Liberto 2015: 120).

A *regime of attraction* is a set of exo-relations that is stable and generates local manifestations in predictable ways (Bryant 2011: 205). For example, gravity, temperature and pressure affect the local manifestations of water on earth. Most surface water we encounter on earth is in a liquid state. Beyond tellurian regimes of attraction, where gravity, temperature and pressure are different, most surface water exists in a solid state (as it does on many moons in the solar system) (Normark 2015a). Hence, the specific regime of attraction for water on earth has been crucial for organic life to emerge and evolve (largely due to the earth's location within the sun's circumstellar habitable zone).

It is not likely that every part of a machine is connected to every other part of the same machine or another machine. However, at least one selective exo-relation must exist for there to be an interaction with another machine. This selection is important in how a *world* emerges. A world consists of 'machines interacting with one another through the mediation of other machines' (Bryant 2014: 113–14). I shall treat the ancient 'Maya culture' as a world where information 'flows' between machines. Whereas we can talk about a 'kingship machine' or an 'agricultural machine' separately, the 'Maya world' emerges from how various parts of each machine are specifically and persistently selected and connected through both human and non-human activities. This process is labelled *worlding* by Descola (2016: 3).

Thus, DeLanda's and Bryant's readings of Deleuze, complexity theories and autopoietic systems emphasise the emergence of machines on multiple scales. This differs from Ingold and several neo-Darwinian cultural evolutionists like Lewens and Shennan who focus on the organism and its immediate relations as an extended phenotype. They seldom treat the evolution of organisations, communities or cities, i.e. machines at larger scales. It is to this that we now shall turn.

## Water, Kingship and Settlement

I shall consider the particular becomings of three multi-scalar machines: water, kingship and settlement in the Prehispanic Maya lowlands in southern Mexico and northern Central America (Figure 6.1), from the Middle Formative Period (600–300 BC) to the late Terminal Classic (AD 950–1100) (Table 6.1). Examples of settlements will be drawn from the Cochuah region.

### *Water*

At its smallest scale, water is a molecule with specific powers, some of which are exemplified below. It can exist in all three states of matter at the same time (imagine ice cubes melting in boiling water). The liquid state of water is also denser than its solid state, which means that ice

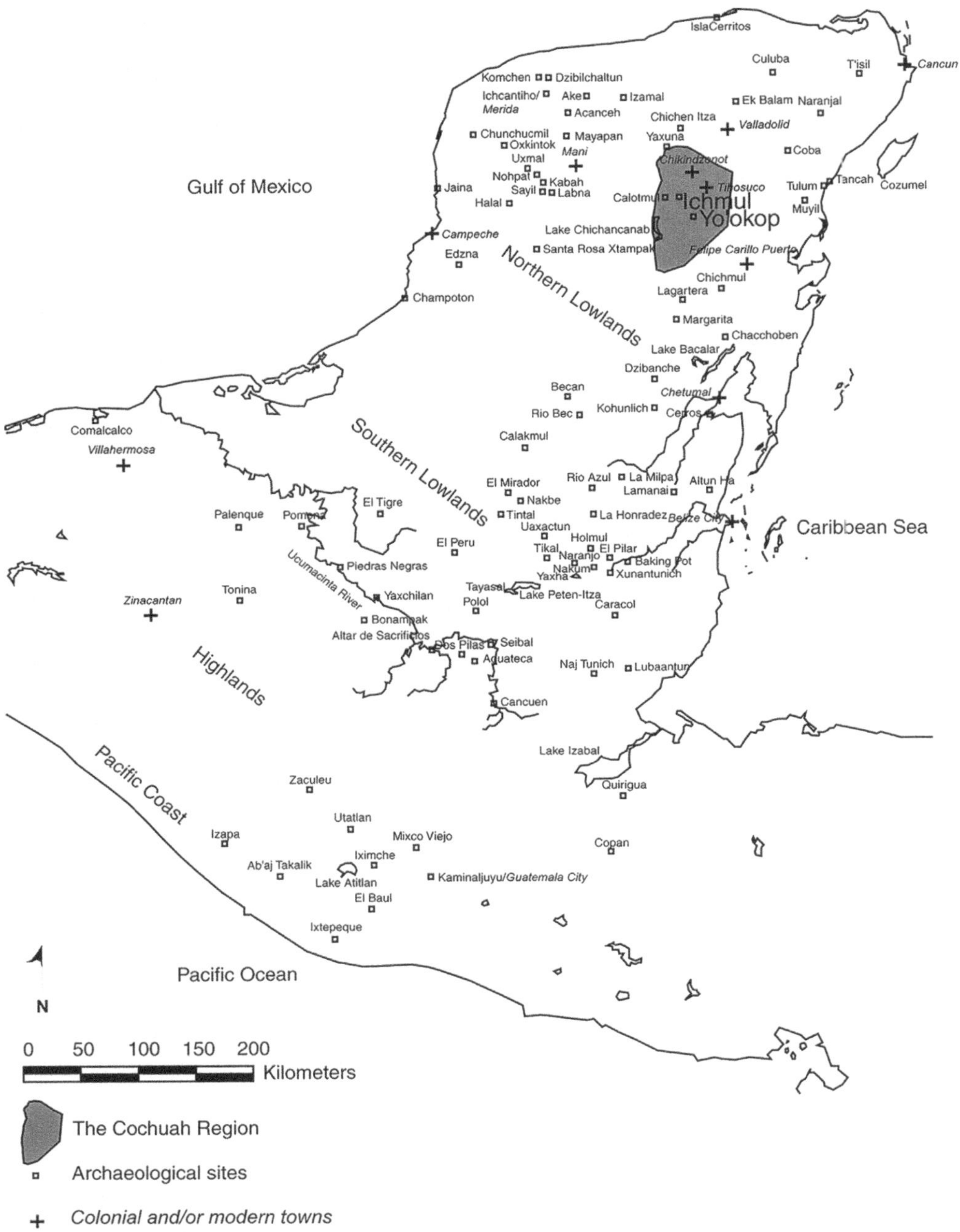


**Figure 6.1** The Maya Area. (Source: Johan Normark with data courtesy of the Cochuah Region Archaeological Survey.)

**Table 6.1** Chronology of the Cochuah Region

| Period | Date |
|---|---|
| Middle Formative | 6/500–300 BC |
| Late Formative | 300 BC–AD 250 |
| Early Classic | AD 250–550 |
| Late Classic | AD 550–750 |
| Terminal Classic | AD 750–1100 |
| Postclassic | AD 1100–1546 |
| Colonial | AD 1546–1810 |
| Historic Mexican | AD 1810–late 20th C |
| Modern | late 20th C–present |

floats in liquid water. Water easily bonds with other molecules important for organisms. These are important powers that allow organisms to evolve on earth, where most water remains in a liquid state that facilitates chemical reactions.

Since liquid water emerged on earth as the planet cooled billions of years ago, it makes sense to begin to think about it not only at the molecular level, but in terms of a global scale. All water on earth can be seen as part of the global *hydrological cycle* (Normark 2014). Here I shall view the hydrological cycle as a vast machine that also crosscuts various 'spheres' (hydrosphere, cryosphere, lithosphere, atmosphere, biosphere, technosphere, noosphere, etc.).

The hydrological cycle is a global machine that is difficult to observe as a whole since we are located within it. This machine is affected by other machines that create a global regime of attraction for organisms on this planet: the sun, the moon, the atmosphere and bodies of water (Normark 2014, 2015a, 2016a). We observe the local manifestations of the hydrological cycle when rain, snow or hail falls, water flows and erodes soil and carves out caves, ice calves, the tide goes in and out, etc.

Humans affect the hydrological cycle directly through damming, diversion, dredging, embanking, draining and irrigation, and indirectly through deforestation and other agricultural activities (Edgeworth 2011).

From this perspective humans have shaped bodies of water and these have shaped us in turn, in a process of niche constructing which I will return to later.

Local water sources in the Maya area fluctuate with the seasons. During the dry season groundwater recedes and droughts are likely to occur, while in the rainy season hurricanes and flooding occur. Regionally this weather pattern is affected by the position of what climate scientists call the Intertropical Convergence Zone (ITCZ), where the trade winds converge. Changes in the location of the ITCZ affect the location of the Hadley Cells, which are the tropical circulation belts directly north and south of the ITCZ that create the trade winds, rain-belts and hurricanes. If the ITCZ moves to the south, the Hadley Cells follow and droughts become more frequent, severe and longer in the Maya area (Gill 2000).

Water is something we constantly encounter since it is manifested by machines on scales between the molecular and the global. For instance, 'invisible' water surrounding the organism, i.e. humidity, affects the respiratory system and the skin. Humidity varies on a daily and annual basis. Too much humidity can lead to irritation and yeast growth on the skin. Too little humidity can cause the skin to dry out and crack. Dry air can also cause asthma attacks (Goad and Gawkrodger 2016; Strauss et al. 1978).

The human body itself consists of water. The exact percentage depends on gender, age and body size. A typical human body consists of roughly 73 per cent water at birth. One year later this drops to around 65 per cent. Through ontogenesis the percentage of water changes in adult bodies. Male adults consist of roughly 60 per cent water. Since fat tissues consist of less water and adult females, on a population level, have more fat than men, they consist of roughly 55 per cent water. Obese people can consist of 45 per cent of water (Guyton 1976). Various body parts contain varying amounts of water. Lungs consist of 84 per cent water, followed by muscles and kidneys (79%), brain and heart (73%), and skin (65%). Bones, common archaeological remains, consist of 32 per cent water (less after death). Teeth contain the least amount of water in the body (5%) (Mitchell et al. 1945: 628).

The ancient Maya noted these bodily differences and incorporated them into their *analogic* ontology. In analogism the world is fragmented and consists of a multiplicity of entities that can be ordered along a scale of degrees because they are not different in kind, such as hot/cold and dry/wet according to a common Mesoamerican analogic system (Descola 2013: 218). For example, Stone (2011) has studied how the female breasts appear in Maya iconography. She argues that the forms of the breasts are indices of the woman's status, as the breasts mirror the woman's life stage affected by changes along the two dualistic poles. Young and fertile women were depicted with larger and rounded breasts whereas post-menopausal women were depicted with sagging breasts. Old women had literally dried out.

A reduction of the volume of bodily fluids may negatively affect the ontogenesis of the human body. Water regulates body temperatures through sweating and respiration. Age, gender and habitat affect the amount of water needed for each individual. Usually, an adult man needs more water than an adult woman. Contemporary use for drinking and cooking in the Maya area amounts to 3.3 litres per day on average (Luzzadder-Beach 2000: 497). McAnany (1990) estimates that each adult Prehispanic individual needed, on average, 4.8 litres on a daily basis (including for drinking, cooking, washing, making ceramics, etc.).

In order to satisfy the organismic need in large-scale communities, water needs to be managed, particularly in an area with an extended dry season where water seldom remains on the karstic surface, as is typical of the Maya landscape and hydroscape. Hence, when the local population of humans (and domestic animals) increases, there is a greater need for *allopoietic machines* (machines produced by other machines [humans] (Bryant 2011, 2014)). Groundwater wells, rain-fed water reservoirs and *chultuns* (subterranean storage chambers) are such machines. Despite the lower human population levels in the colonial period compared to the Terminal Classic period, the need for wells increased because livestock was introduced (Normark 2016a).

One problem with water sources in the Maya area is that standing water can become stagnant and generate conditions for the survival and propagation of insects and parasites. It can also create noxious chemicals, such as nitrogen (Lucero 2002: 815). This water would have been boiled or combined with fermented maize gruel to satisfy thirst (Scarborough et al. 2012: 12412). Water, combined with other substances, enters and transforms the organism's metabolism. You are what you eat and drink. Indeed, according to the Kiché Maya creation account, Popol Vuh, the first true humans were made from maize gruel (Tedlock 1996).

With the emergence of allopoietic water machines the landscape and waterscape also became a new *taskscape*, i.e. a set of connected activities (Ingold 1993). In most of the pre-industrial world the management of water was a local development. Scarborough (2009) makes a distinction between *technotasking* and *labortasking* within the Maya landscape and waterscape. The first was introduced to the Maya area by the Spaniards in the sixteenth century. It is linked to an expansionist logic that often exploits resources rapidly. Labortasking, which characterised the Prehispanic Maya, creates built enhancements (such as reservoirs and architectural features that collect water), but it leads to social complexity at a lower rate of change than during technotasking. Labortasking resource-specialised community organisations existed throughout the Prehispanic Maya lowlands (Scarborough 2009: 202). The difference between these two strategies can be seen in the Cochuah region, where large colonial period sites have far more wells than the Prehispanic sites of the same size. This is due to imported cattle and a new exploitative economy that created a new taskscape (Normark 2016a).

## *Kingship*

The Classic period 'divine kingship' (*ajawlel*) is the most visible 'institution' in the ancient Maya world (Martin and Grube 2000). As an organisation, kingship and the royal court are often associated with the origins of the state in orthogenetic evolutionary theory (Johnson 2010). Water management on a grand scale precedes the earliest known traces of kingship in the Maya lowlands but this management created the basis that sustained kingship. The development of water-management machines intensified institutional development. Kingship emerged, evolved and dissolved partly in relation to water management. Lucero et al. state that 'the Maya royal house of cards collapsed because it relied largely on water control' (2011: 487). Once in existence, kingship evolved and changed until its demise with the 'Maya collapse' in the ninth to eleventh centuries. Its remaining parts became new hydrosocial machines.

In Deleuzo-Guattarian terms, kingship was an *apparatus of capture* that overcoded the presignifying regime of signs associated with water. In this case the presignifying regime was animistic as defined by Descola (2013). *Ajawlel* overcoded previous animistic meanings of water by, among other things, appropriating community rituals. Water became associated with the signifying regime of signs centred upon the ruler (Deleuze and Guattari 1987; Normark 2012, 2015b). The signifying regime of signs loosely resembles Descola's (2013) definition of analogism, in which autonomous entities are recombined into networks by analogies. A common analogic network in the Maya area is the quadripartite pattern with four corners or sides which can be found in many archaeological assemblages, such as settlement layout (buildings surrounding a square or rectangular plaza). The body of the ruler also referred to this network as did the very structure of the world (Normark in prep.).

Through this overcoding, water worked as a medium that extended the domains of kingship. Water in manageable situations acted as a sorting and selective device, particularly during droughts which forced the population across a larger area to locations where water sources could be maintained (Normark 2016a).

## *Settlement*

The hydrological cycle was and still is *global* in scale whereas *ajawlel* was *regional*. At each *local* archaeological site from the Late Formative and Classic periods the emergent settlement was the result of exo-relations between the hydrological cycle, kingship and other machines too numerous to list but which include soils, trade items, celestial objects (aligning buildings according to the seasonal positions of planets and stars), and so on. Hence, water actualised both repetitive and unique settlement patterns. In this process water sources became meaningful because landscapes and waterscapes 'are meaningful for human purposes when water is not *everywhere*, but rather *somewhere*, as there is no meaning without difference' (Mangiameli 2013: 148, original emphasis).

A settlement with its allopoietic water machine(s) can be understood through *niche construction theory* in which the local ecology is modified by organisms and organisms are modified by the niches of other organisms. Hence, niches emerge and disappear through the workings of organisms (Fuentes 2013: 50). The development and growth of an individual organism depends on its structural coupling with other entities (Ramirez-Goicoechea 2013: 69). Within an *ontogenetic niche* the behaviour of organisms emerges through active processes within multiscalar machines that include 'molecular, cellular, ecological, social and political experience, together with other developmental resources' (71). Kingship and allopoietic water machines participated in creating these ontogenetic niches in the Maya world.

The niches produced multiple manifestations of settlements. One manifestation was settlement centralisation to one site core, such as Ichmul, where a narrow funnel-shaped sinkhole (*cenote*) probably became the centre of the site, with the royal or elite compounds located to the south and east (Figure 6.2). A second was centralisation to several cores, such as Yo'okop, where the water reservoir or lake lies in between the two largest groups (Figure 6.3). A third one has no monumental site core, such as Nohcacab, located in between Ichmul and Yo'okop (Figure 6.4). Here water was obtained from one artificial well, not located at the largest house compound, and, perhaps significantly, kingship was only indirectly present (Normark 2009, 2010, 2018).

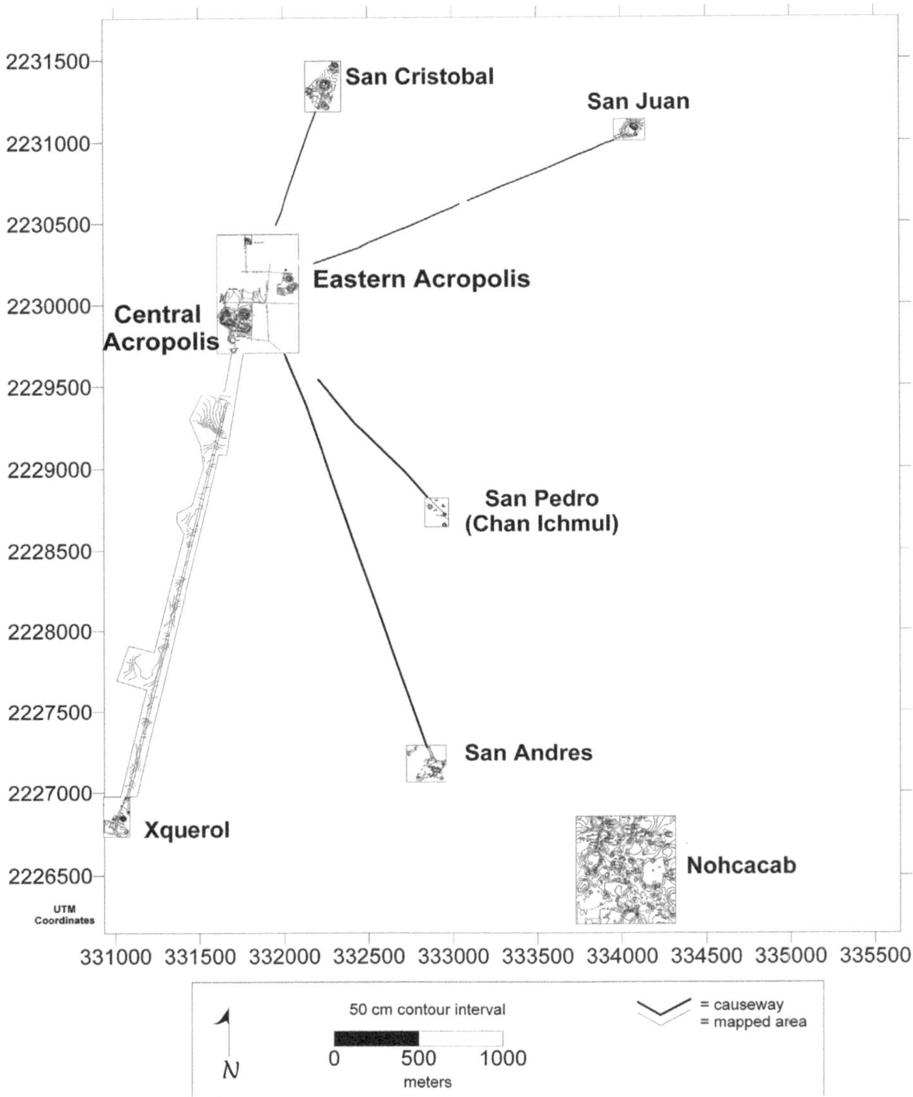


**Figure 6.2** Ichmul. (Source: Johan Normark with data courtesy of the Cochuah Region Archaeological Survey.)

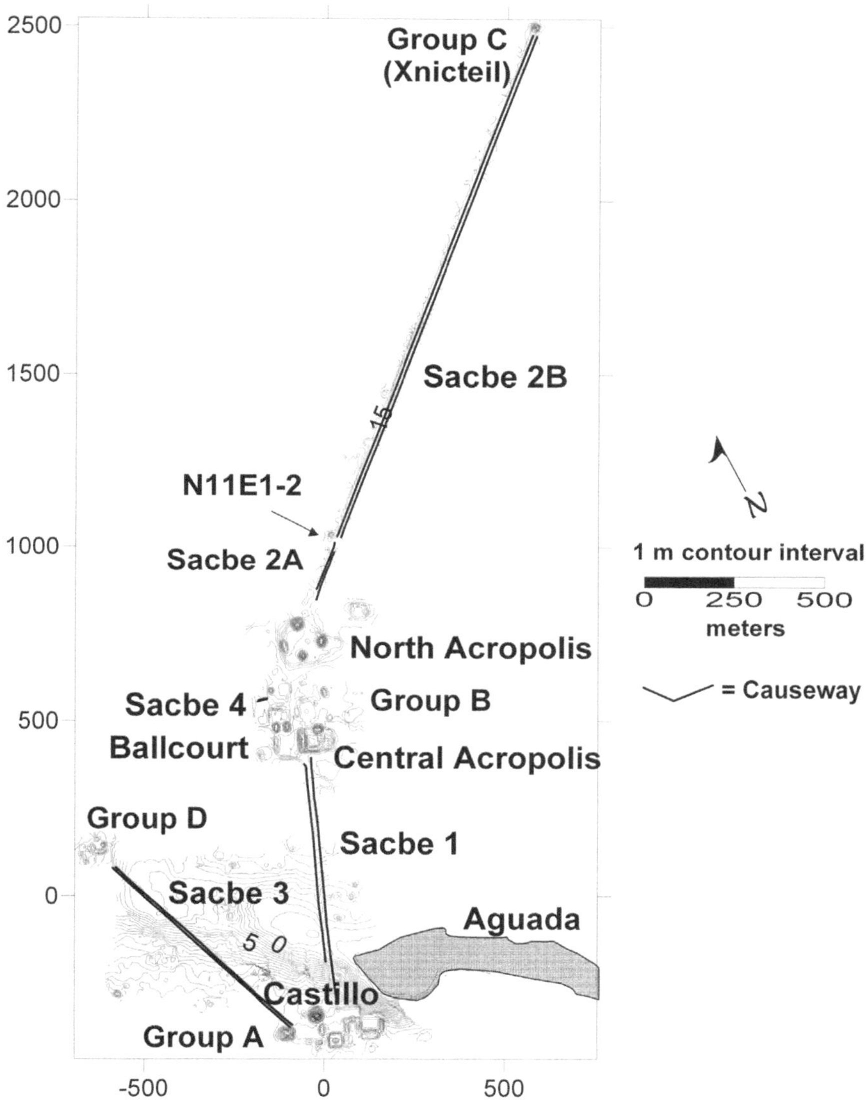


Figure 6.3 Yo'okop. (Source: Johan Normark with data courtesy of the Cochuah Region Archaeological Survey.)

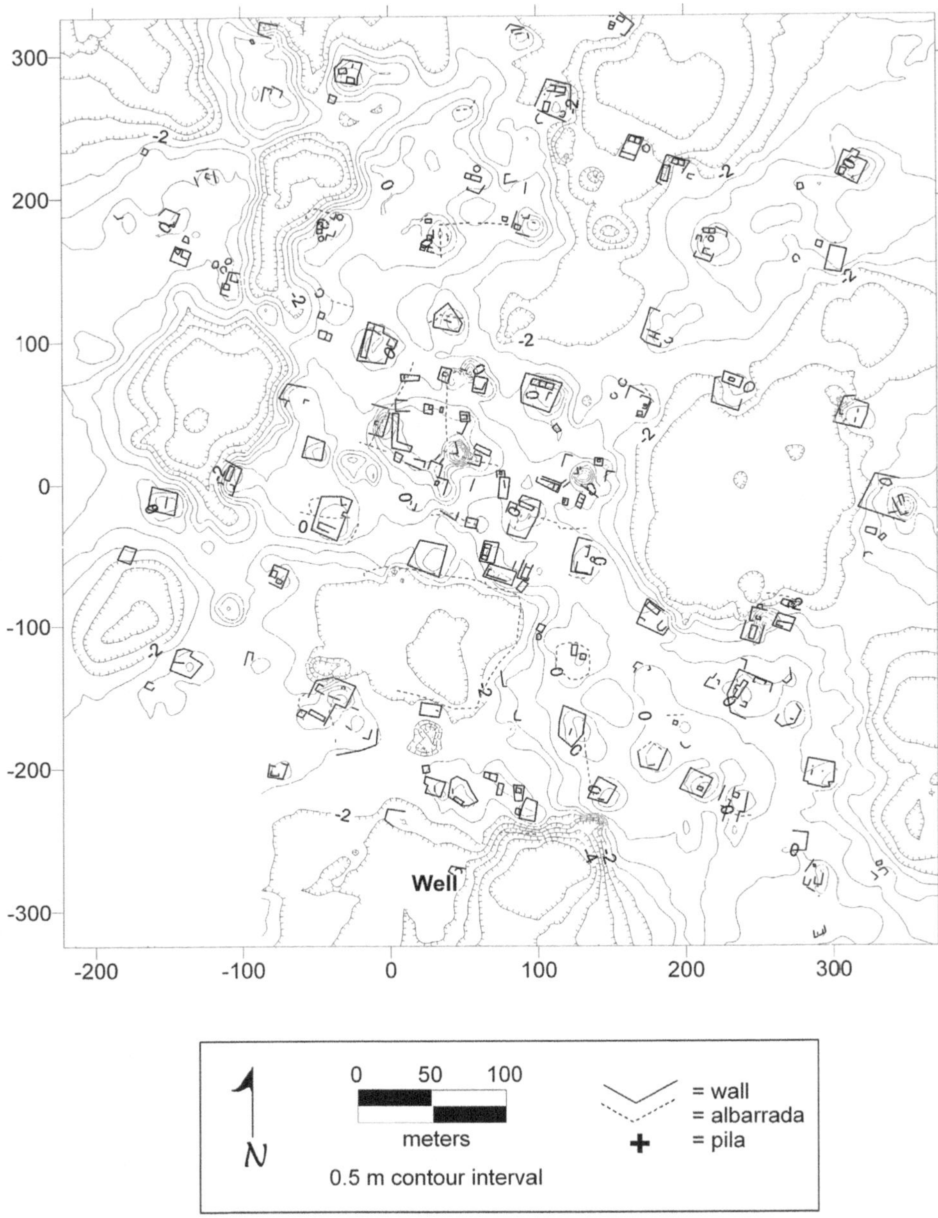


**Figure 6.4** Nohcacab. (Source: Johan Normark with data courtesy of the Cochuah Region Archaeological Survey.)

Maya settlements were differentiated in terms of people's access to resources which were organised in relation to either residence or kingship (Gillespie 2000). These organisations selected and sorted the population into structured relations. Most 'of the ranking or sorting processes that maintain the differential access to economic and cultural capital are *resource dependence* relations that exist not between people but between institutional organizations' (DeLanda 2006: 65, emphasis original). It is often argued that the royal elite controlled water sources that were needed by kinship groups (Lucero 2006; Scarborough 1998). This is the case at larger sites centred around rain-fed water reservoirs and cenotes. However, smaller water sources not in the control of an elite were also common (Normark 2016a; Weiss-Krejci and Sabbas 2002).

Settlements were generated and affected by construction activities, political decisions, trade routes, topography, weather patterns and so on. As some sites grew in size, they also transformed their niches. Early on in their development both Yo'okop and Ichmul may have been similar to Nohcacab, but their niches sent them along different trajectories. Nohcacab's growth was constrained by nearby Ichmul during the Early Classic. Few archaeological remains at Nohcacab date to the Early Classic compared to the previous Late Formative. However, Nohcacab also benefited from Ichmul's settlement expansion during the Terminal Classic (Normark 2018).

Another way to view how these niches affected each other is through Bryant's concept of *gravitational fields* (2014: 196). The gravitational fields of machines affect the becomings of other machines. Some machines are either captured in the 'orbit' of other machines or the relation is more symmetrical. This gravity is a field or topology which is followed by other machines during their movement (186). A *bright machine* gravitationally overcodes the movements of other machines. The 'water machines' were such bright machines since they attracted settlement. Machines that end up in the orbit of a bright machine become *satellites* (202–3). Satellites are found on different spatiotemporal scales. People living at Nohcacab were satellites to the well at the site. Nohcacab as a settlement became a satellite to Ichmul and remained so for generations. Thus, most machines are satellites at some scale, or in some context, and there may be few ways to break out of orbits caused by bright machines.

## Discussion and Conclusion

It was on the scale of the individual human organism that kingship and water intersected in the settlements. The Classic period Maya individual was a hierarchically ordered body within the *ajawlel* machine (Normark 2012). Affective cognition was regulated along political lines and these powers were therefore differentially distributed within the population (Protevi 2009: 29–30), following the hierarchical stratification prevalent in Classic Maya culture. Archaeologically speaking, these distributions can be seen at various locations (temple, causeway, water reservoir, palace, house, cave, etc.), and in groups and organisations (extended family, lineage, kingship, etc.).

Water and kingship differentiated people and settlement through various sorting and selecting mechanisms. Rituals were such mechanisms since many of the communal events people participated in were ritualised and ceremonialised. These rituals were selective ways of binding people to a group or excluding others (Kertzer 1988; Munson et al. 2016). Water-related rituals were also machines in themselves as they were repeated for centuries and extended across large areas. The ritual life of a ruler was therefore often affected by and entrained with the hydrological cycle.

Greater annual and periodic rituals and ceremonies were likely held in processions between various parts of the settlements. Processions on causeways (*sacbeob*) physically manifested the local domains of the ruler. Yo'okop (Figure 6.3) lacks a central core. Instead there were multiple cores connected by causeways that may have had varying importance at different periods or reflected different houses or lineages. Terminal Classic Ichmul (Figure 6.2) connected previously peripheral sites into a larger centralised machine through a substantial causeway construction project (Flores Colin 2015; Normark 2008, 2010).

In the Maya area, it is often argued, the ruler ideologically controlled water, rain, maize and so on. (Lucero 2006; Scarborough 1998). Even though ideologies and discourses can be labelled as bright machines in their own right, humans are also affected by the physical aspects of machines. To some extent the Classic Maya kings were satellites to both their allopoietic water machines and domesticated maize. Maize is therefore another example of a bright machine in the Maya area. It has a seasonal rhythm that affected much of Maya life. During the Classic period common toponyms included a morpheme (*-nal*) which refers to corn plants. Settlements were probably viewed as cornfields or places of corn plants (Tokovinine 2013: 9). Local varieties of maize gods seem to reflect an emic perspective of group identities and classification (123). Just as Pollan (2002) argues that potatoes exercised natural selection on humans during the Great Irish Potato Famine, one can argue that maize selected Classic Maya kingship for extinction because the Classic Maya relied too much on maize and water reservoirs (Webster 2014).

The 'Maya collapse' came fairly unexpectedly even if the process dragged on in some areas (Ebert et al. 2014). This partly fits the description of Bryant's *rogue* machine which is not connected to the gravitational field of other machines. Either it unexpectedly comes from a different location or it emerges locally as the result of something unexpected that reconfigures the gravitational field (Bryant 2014: 208f.). Examples of rogue machines are the Bubonic plague, the 9/11 terrorist attacks, the 2004 tsunami, the Chicxulub impact that ended the late Cretaceous period, the Spanish conquest of Mexico, and the severe droughts alleged to have caused the Maya kingship to collapse. In the latter case, there are few contemporary textual sources mentioning drastic events, since droughts were a recurring problem. However, the severity of the droughts was intensified by increased demographic pressure and diminishing returns (Tainter 1995). The rogue machine is therefore similar to a Black Swan event in Taleb's (2007) discussion of highly improbable events. When faced with the Black Swan, kingship was fragile compared to the livelihood of the forest-garden farmers, since the latter continues today (Ford and Nigh 2015; Ford et al. 2001).

No single machine caused the collapse(s). There were several that shattered the regime of attraction of Maya kingship, such as drought, trade route changes, and so on (Demarest 2014; Iannone 2014). However, droughts were more crucial than other proposed causes. Without the hierarchical kingship, farmers had no need to remain in larger communities and dispersion began. Put in Deleuzo-Guattarian terminology, without kingship as an apparatus of capture, the farming communities *deterritorialised* and *reterritorialised* elsewhere as new hydrosocial becomings. Hence, the 'Maya collapse' involved the disappearance of some machines, but not of the Maya world.

In sum, by introducing a Deleuzo-Guattarian vocabulary into studies on the significance of water for Maya culture and the collapse of Maya kingship, this chapter has demonstrated how a multi-scalar approach to social formations can rehabilitate the idea of cultural evolution and separate it from both the orthogenetic evolutionism that used to dominate the field of archaeology and the neo-Darwinism that is dominant today.

## Note

1. This chapter is the result of the project 'Water as an Archaeological Material', financed by Riksbankens Jubileumsfond (P10–0367:1). I have used data collected under Justine Shaw's and Dave Johnstone's INAH permit for fieldwork in the Cochuah region (2000–5, 2008, 2010, 2012 and 2014). Their fieldwork has been financed by Antigua Foundation, the Foundation for the Advancement of Mesoamerican Studies, Inc., the H. John Heinz III Fund, the Selz Foundation.

## References

Ansell-Pearson, K. (1999), *Germinal Life: The Difference and Repetition of Deleuze*, London and New York: Routledge.

Beloussov, L. (1997), 'Life of Alexander G. Gurwitsch and His Relevant Contribution to the Theory of Morphogenetic Fields', *International Journal of Developmental Biology* 41: 6, pp. 771–7.

Bryant, L. R. (2011), *The Democracy of Objects*, Ann Arbor: Open Humanities Press.

Bryant, L. R. (2014), *Onto-Cartography: An Ontology of Machines and Media*, Edinburgh: Edinburgh University Press.

Clark, A. (2003), *Natural Born Cyborgs: Minds, Technologies and the Future of Human Intelligence*, Oxford: Oxford University Press.

DeLanda, M. (1997), *A Thousand Years of Nonlinear History*, New York: Zone.

DeLanda, M. (2002), *Intensive Science and Virtual Philosophy*, London: Continuum.

DeLanda, M. (2006), *A New Philosophy of Society: Assemblage Theory and Social Complexity*, London: Continuum.

DeLanda, M. (2011), *Philosophy and Simulation: The Emergence of Synthetic Reason*, London: Bloomsbury Academic.

Deleuze, G. and F. Guattari (1987), *A Thousand Plateaus: Capitalism and Schizophrenia*, trans. B. Massumi, Minneapolis: University of Minnesota Press.

Demarest, A. A. (2014), 'The Classic Maya Collapse, Water, and Economic Change in Mesoamerica', *The Great Maya Droughts in Cultural Context: Case Studies in Resilience and Vulnerability*, ed. G. Iannone, Boulder, CO: University Press of Colorado, pp. 177–206.

de Queiroz, K. (2005), 'Ernst Mayr and the Modern Concept of Species', *Proceedings of the National Academy of Sciences* 102, pp. 6600–7.

Descola, P. (2013), *Beyond Nature and Culture*, Chicago: University of Chicago Press.

Descola, P. (2016), 'Biolatry: A Surrender of Understanding (Response to Ingold's "A Naturalist Abroad in the Museum of Ontology")', *Anthropological Forum* 26: 3, pp. 321–8.

Di Liberto, Y. (2015), 'Machines, Virtual and Attractors: Levi Bryant's Machine-Oriented Ontology and Substances as Powers', *Methode* 4: 5, pp. 112–32.

Ebert, C. E., K. M. Prufer, M. J. Macri, B. Winterhalder and D. J. Kennett (2014), 'Terminal Long Count Dates and the Disintegration of Classic Period Maya Polities', *Ancient Mesoamerica* 25: 2, pp. 337–56.

Edgeworth, M. (2011), *Fluid Pasts: Archaeology of Flow*, Bristol: Bristol Classical Press.

Finlayson, C. (2014), *The Improbable Primate: How Water Shaped Human Evolution*, Oxford: Oxford University Press.

Flores Colin, A. (2015), 'Artificial Causeways: Metaphors of Power and Death', *The Maya of the Cochuah Region: Archaeological and Ethnographic Perspectives on the Northern Lowlands*, ed. J. M. Shaw, Albuquerque: University of New Mexico Press, pp. 133–56.

Ford, A., R. Larios, J. Normark, P. Morales and C. Ramos (2001), *Influence of Ancient Settlement in the Contemporary Maya Forest: Investigating Land Use at El Pilar*, Santa Barbara: Mesoamerican Research Center, University of California.

Ford, A. and R. Nigh (2015), *The Maya Forest Garden: Eight Millennia of Sustainable Cultivation of the Tropical Woodlands*, Walnut Creek: Left Coast Press.

Foucault, M. (1969), *L'archéologie du savoir*, Paris: Gallimard.

Fuentes, A. (2013), 'Blurring the Biological and Social in Human Becomings', *Biosocial Becomings: Integrating Social and Biological Anthropology*, ed. T. Ingold and G. Palsson, Cambridge: Cambridge University Press, pp. 42–58.

Gill, R. B. (2000), *The Great Maya Droughts*, Albuquerque: University of New Mexico Press.

Gillespie, S. D. (2000), 'Rethinking Ancient Maya Social Organization: Replacing "Lineage" with "House"', *American Anthropologist* 102, pp. 467–84.

Goad, N. and D. J. Gawkrodger (2016), 'Ambient Humidity and the Skin: The Impact of Air Humidity in Healthy and Diseased States', *Journal of the European Academy of Dermatology and Venereology* 30: 8, pp. 1285–94.

Guyton, A. C. (1976), *Textbook of Medical Physiology*, 5th edn, Philadelphia: W. B. Saunders.

Hamilakis, Y. (2013), *Archaeology and the Senses: Human Experience, Memory, and Affect*, Cambridge: Cambridge University Press.

Harris, O. J. T. (2014), '(Re)assembling Communities', *Journal of Archaeological Method and Theory* 21, pp. 76–97.

Iannone, G. (ed.) (2014), *The Great Maya Droughts in Cultural Context: Case Studies in Resilience and Vulnerability*, Boulder, CO: University Press of Colorado.

Ingold, T. (1993), 'The Temporality of the Landscape', *World Archaeology* 25: 2, pp. 152–74.

Ingold, T. (2000), *The Perception of the Environment: Essays in Livelihood, Dwelling and Skill*, London: Routledge.

Ingold, T. (2013), 'Prospect', *Biosocial Becomings: Integrating Social and Biological Anthropology*, ed. Tim Ingold and Gisli Palsson, Cambridge: Cambridge University Press, pp. 1–21.

Ingold, T. and G. Palsson (eds) (2013), *Biosocial Becomings: Integrating Social and Biological Anthropology*, Cambridge: Cambridge University Press.

Johnson, M. (2010), *Archaeological Theory: An Introduction*, Oxford: Wiley-Blackwell.

Kertzer, D. I. (1988), *Ritual, Politics and Power*, New Haven: Yale University Press.

Lewens, T. (2015), *Cultural Evolution: Conceptual Challenges*, Oxford: Oxford University Press.

Lucas, G. (2012), *Understanding the Archaeological Record*, Cambridge: Cambridge University Press.

Lucero, L. J. (2002), 'The Collapse of the Classic Maya: A Case for the Role of Water Control', *American Anthropologist* 104: 3, pp. 814–26.

Lucero, L. J. (2006), *Water and Ritual: The Rise and Fall of Classic Maya Rulers*, Austin: University of Texas Press.

Lucero, L. J., J. D. Gunn and V. L. Scarborough (2011), 'Climate Change and Classic Maya Water Management', *Water* 3, pp. 479–94.

Luzzadder-Beach, S. (2000), 'Water Resources of the Chunchucmil Maya', *The Geographical Review* 90: 4, pp. 493–510.

Lyman, R. L. and M. J. O'Brien (2001), 'The Direct Historical Approach, Analogical Reasoning, and Theory in Americanist Archaeology', *Journal of Archaeological Method and Theory* 8: 4, pp. 303–42.

McAnany, P. A. (1990), 'Water Storage in the Puuc Region of the Northern Maya Lowlands: A Key to Population Estimates and Architectural Variability', *Precolumbian Population History in the Maya Lowlands*, ed. T. P. Culbert and D. S. Rice, Albuquerque: University of New Mexico Press, pp. 263–84.

Malafouris, L. (2013), *How Things Shape the Mind: A Theory of Material Engagement*, Cambridge, MA: MIT Press.

Mangiameli, G. (2013), 'The Habits of Water: Marginality and the Sacralization of Non-Humans in North-Eastern Ghana', *Biosocial Becomings: Integrating Social and Biological Anthropology*, ed. T. Ingold and G. Palsson, Cambridge: Cambridge University Press, pp. 145–61.

Martin, S. and N. Grube (2000), *Chronicle of Maya Kings and Queens*, London: Thames and Hudson.

Mitchell, H. H., T. S. Hamilton, F. R. Steggerda and H. W. Bean (1945), 'The Chemical Composition of the Adult Human Body and Its Bearing on the Biochemistry of Growth', *Journal of Biological Chemistry* 158, pp. 625–37.

Munson, J., J. Scholnick, M. Looper, Y. Polyukhovych and M. J. Macri (2016), 'Ritual Diversity and Divergence of Classic Maya Dynastic Traditions: A Lexical Perspective on Within-Group Cultural Variation', *Latin American Antiquity* 27: 1, pp. 74–95.

Neimanis, A. (2012), 'Hydrofeminism: Or, on Becoming a Body of Water', *Undutiful Daughters: Mobilizing Future Concepts, Bodies and Subjectivities in Feminist Thought and Practice*, ed. H. Gunkel, C. Nigianni and F. Söderbäck, New York: Palgrave Macmillan.

Normark, J. (2006), *The Roads In-Between: Causeways and Polyagentive Networks at Ichmul and Yo'okop, Cochuah Region, Mexico*, Dissertation thesis, Göteborg: University of Gothenburg.

Normark, J. (2008), 'The Triadic Causeways of Ichmul: Virtual Highways Becoming Actual Roads', *Cambridge Archaeological Journal* 18: 2, pp. 215–37.

Normark, J. (2009), 'The Making of a Home: Assembling Houses at Nohcacab, Mexico', *World Archaeology* 41: 3, pp. 430–44.

Normark, J. (2010), 'Involutions of Materiality: Operationalizing a Neo-materialist Perspective Through the Causeways of Ichmul and Yo'okop, Mexico', *Journal of Archaeological Method and Theory* 17: 2, pp. 132–73.

Normark, J. (2012), 'The Road of Life: Bodies-Politic in the Maya Area', *To Tender Gender: The Pasts and Futures of Gender Research in Archaeology*, ed. I.-M. Back-Danielsson and S. Thedéen, Stockholm: Stockholm University, pp. 117–36.

Normark, J. (2014), 'Water as a Hyperfact', *Current Swedish Archaeology* 22, pp. 155–78.

Normark, J. (2015a), 'Going Against the Flow: Reaction to Veronica Strang', *Archaeological Dialogues* 22: 2, pp. 199–206.

Normark, J. (2015b), 'Not Only the Home of the Earth Lord: Cochuah Caves as Holy Places, Holey Spaces, and Emergent Wholes', *The Maya of the Cochuah Region: Archaeological and Ethnographic Perspectives on the Northern Lowlands*, ed. J. M. Shaw, Albuquerque: University of New Mexico Press, pp. 171–93.

Normark, J. (2016a), 'The Chicxulub Impact and Its Different Hydrogeological Effects on Prehispanic and Colonial Settlement in the Yucatan Peninsula', *WIREs Water* 3, pp. 871–84.

Normark, J. (2016b), 'Multi-scalar Cognitive Time: Experiential Time, Known Time and Maya Calendars', *Quaternary International* 405, pp. 52–60.

Normark, J. (2018), 'Settlement Topology: The Rotted Community and the Congregated Community in the Cochuah Region, Yucatan, Mexico', *Urban Variation: Utopia, Planning and Practice*, ed. P. Cornell, L. Ersgård and A. Nilsen, Lulu Press, pp. 751–98.

Normark, J. (in prep.), 'Analogic Maya Caves: Descola's Modes of Identification in a Mesoamerican Context'.

Palsson, G. (2013), 'Ensembles of Biosocial Relations', *Biosocial Becomings: Integrating Social and Biological Anthropology*, ed. T. Ingold and G. Palsson, Cambridge: Cambridge University Press, pp. 22–41.

Pollan, M. (2002), *The Botany of Desire: A Plant's Eye View of the World*, New York: Random House.

Protevi, J. (2009), *Political Affect: Connecting the Social and the Somatic*, Minneapolis: University of Minnesota Press.

Ramirez-Goicoechea, E. (2013), 'Life-in-the-Making: Epigenesis, Biocultural Environments and Human Becomings', *Biosocial Becomings: Integrating Social and Biological Anthropology*, ed. T. Ingold and G. Palsson, Cambridge: Cambridge University Press, pp. 59–83.

Richerson, P. J. and R. Boyd (2005), *Not by Genes Alone: How Culture Transformed Human Evolution*, Chicago: University of Chicago Press.

Scarborough, V. L. (1998), 'Ecology and Ritual: Water Management and the Maya', *Latin American Antiquity* 9: 2, pp. 135–59.

Scarborough, V. L. (2009), 'The Archaeology of Sustainability, Mesoamerica', *Ancient Mesoamerica* 20: 2, pp. 197–203.

Scarborough, V. L., N. P. Dunning, K. B. Tankersley, C. Carr, E. Weaver, L. Grazioso, B. Lane, J. G. Jones, P. Buttles, F. Valdez and D. L. Lentz (2012), 'Water and Sustainable Land Use at the Ancient Tropical City of Tikal, Guatemala', *Proceedings of the National Academy of Sciences* 109: 31, pp. 12408–13.

Shanks, M. and C. Tilley (1987), *Social Theory and Archaeology*, Cambridge: Polity Press.

Shennan, S. (2002), *Genes, Memes and Human History: Darwinian Archaeology and Cultural Evolution*, London: Thames and Hudson.

Stone, A. J. (2011), 'Keeping Abreast of the Maya: A Study of the Female Body in Maya Art', *Ancient Mesoamerica* 22: 1, pp. 167–83.

Strang, V. (2014), 'Fluid Consistencies: Material Relationality in Human Engagements with Water', *Archaeological Dialogues* 21: 2, pp. 133–50.

Strauss, R. H., E. R. McFadden, R. H. Ingram, E. C. Deal and J. J. Jaeger (1978), 'Influence of Heat and Humidity on the Airway Obstruction Induced by Exercise in Asthma', *The Journal of Clinical Investigation* 61, pp. 433–40.

Tainter, J. A. (1995), 'Sustainability of Complex Societies', *Futures* 27, pp. 397–407.

Taleb, N. N. (2007), *The Black Swan: The Impact of the Highly Improbable*, London: Allen Lane.

Tedlock, D. (1996), *Popol Vuh: The Definite Edition of the Mayan Book of the Dawn of Life and the Glories of Gods and Kings*, New York: Simon and Schuster.

Tokovinine, A. (2013), *Place and Identity in Classic Maya Narratives*, Washington: Dumbarton Oaks Research Library and Collection.

Webster, D. (2014), 'Maya Drought and Niche Inheritance', *The Great Maya Droughts in Cultural Context: Case Studies in Resilience and Vulnerability*, ed. G. Iannone, Boulder, CO: University Press of Colorado, pp. 333–58.

Weiss-Krejci, E. and T. Sabbas (2002), 'The Potential Role of Small Depressions as Water Storage Features in the Central Maya Lowlands', *Latin American Antiquity* 13: 3, pp. 343–57.

Chapter 7

# Against Social Evolution: Deleuze and Guattari's Social Topology

*Daniel W. Smith*

*Complex states did not and could not have evolved out of more 'primitive' hunter-gatherer societies.* This is the profound thesis that lies at the heart of Deleuze and Guattari's critique of traditional theories of social evolution. The widespread presumption that human societies evolved progressively from the simple to the complex, from the 'primitive' to the 'civilised', from hunter-gatherer groups to large state formations, received perhaps its paradigmatic formulation in Lewis Henry Morgan's 1877 book, *Ancient Society; Or: Researches in the Lines of Human Progress from Savagery through Barbarism to Civilization*, a work that had a profound influence on numerous nineteenth-century thinkers, most notably Marx and Engels (Morgan 1877; Carneiro 2003). The title of the third chapter of Deleuze and Guattari's *Anti-Oedipus* ('Savages, Barbarians, Civilized Men') is derived from Morgan's work, even though Morgan's name is never mentioned (Deleuze and Guattari 2009: 139). But the reference to Morgan's linear and progressive concept of social evolution is clearly meant to be provocative, since the universal history developed in the two volumes of *Capitalism and Schizophrenia* is not only explicitly directed *against* conceptions of social progress, but is grounded in what Deleuze calls a 'non-chronological' conception of time (Deleuze 1989: 79).[1] Time as *succession* gives way to time as *coexistence*: time does not move from one actual moment to another (chronology), but rather from the virtual to the actual (actualisation).[2] Deleuze and Guattari are not denying social change, but they are arguing that we cannot understand social change unless we see it as taking place within a *field of coexistence*.

Deleuze frequently noted that his concept of time was initially derived from biology, and especially embryology (Ruyer): the apparent chronology of a life in terms of a past, a present and a future is in fact the unfolding of the potential of an egg (a body *without* organs), an unceasing genetic movement from the virtual to the actual.[3] Deleuze's metaphysics of time is in a sense a generalisation of this biological fact: time is no longer a measure of movement, as it was for the Greeks; rather, all movement – whether cosmic, biological or social – must be understood to be unfolding synthetically within a topological field. *Capitalism and Schizophrenia* thus not only contains one of Deleuze and Guattari's most sustained engagements with evolutionary theory, but it also provides one of the most concrete examples of how to analyse the 'evolution' of a domain (in this case, the socio-political domain) from the viewpoint of a temporal field of coexistence.

## The State Had No 'Origin'

Deleuze and Guattari's critique of social evolution begins with the theory of the State, and a consideration of the nature of ancient despotic States such as Sumer, Babylon or Egypt. What was the origin of such States, and how did they acquire their astonishing dominance? Marx suggested a famous answer to such questions: the archaic State was a milieu of interiority that managed to stockpile the surplus production of the surrounding agricultural communities ('primitive accumulation'), thereby constituting a transcendent public power that converged on the person of the despot (Marx 1965: 69–70; cf. Deleuze and Guattari 2009: 194). Following Marx, the great archaeologist V. Gordon Childe proposed a theory of the origins of prehistoric states that has become canonical (Childe 1951, 2009; cf. Lull and Mico 2011: 180–9): at some point in prehistory, hunter-gatherer groups learned to cultivate grain and raise livestock (the Neolithic Revolution), and it was the surplus of agricultural food that is supposed to have made the State possible, with its complex divisions of labour, large economic projects and intricate social organisation (the Urban Revolution) (Deleuze and Guattari 1987: 428). In other words, primitive societies eventually reached a threshold in their 'mode of production' that allowed them to pass from an economy of *subsistence* to an economy of *surplus*. Using two complementary arguments, Deleuze and Guattari contend that the evidence from archaeology, ethnography and even history does not support this theory.

The first argument comes from the analysis of primitive societies. Pierre Clastres, in his 1974 book *Society Against the State* (Clastres 1989; cf. Clastres 1994) had shown that the absence of a State in primitive societies is not a sign that they were 'backward' societies that had not yet evolved or developed enough. On the contrary, primitive societies are constituted by mechanisms that deliberately *ward off* the apparatus of the State, and actively prevent it from appearing. Clastres emphasised two such mechanisms: the role of *chiefs*, whose status constantly waxes and wanes, thereby preventing the resonance of power in a single despot; and the function of *war*, which maintained polemical relations of antagonism between segmentary lineages, preventing their convergence in a state apparatus. Clastres had been influenced, in part, by Marshal Sahlins' *Stone Age Economics*, which argued that hunter-gatherers, far from living at a subsistence level requiring constant toil, were in fact the first affluent society, where the quest for food was intermittent and leisure was abundant (Sahlins 1972; see Clastres 1994). The absence of a surplus did not indicate an inability to develop technical means or overcome environmental obstacles, but was a positive goal, socially valorised. Even the innovations imported by colonialists were utilised, not to increase production but to reduce work time. The work of both Clastres and Sahlins, in turn, had been anticipated by Marcel Mauss, whose 1925 essay *The Gift* had already shown that the giving of gifts and counter-gifts (potlatch) in primitive societies was a mechanism for warding off the accumulation of wealth (Mauss 1954: 3). In short, numerous anthropologists have identified positive mechanisms in primitive societies that actively *prevent* the formation of a State apparatus: there is a refusal of the State's apparatus of power as much as a refusal of markets and the economy. Primitive societies, in this sense, are 'self-validating' (Deleuze and Guattari 2009: 203). If this claim is correct, it makes the appearance of the State difficult to explain: How could the State have evolved out of primitive hunter-gatherer societies if these are societies whose very organisation is directed *against* the formation of the State (Silbertin-Blanc 2013: 22)?

The second argument comes from the analysis of the State. The urbanist Jane Jacobs, in the first chapter of her 1969 book *The Economy of Cities*, entitled 'Cities First–Rural Development Later', launched an attack on what she called 'the dogma of agricultural primacy' (1970: 3), the idea that an agricultural surplus was the condition for the appearance of the State. Jacobs, contentiously, attempted to invert this schema: it is the State that creates agriculture, she argued, and not the converse. She based her conclusions in part on James Mellaart's discovery of Çatalhöyük, a 'proto-town' in Turkey that dates back to Neolithic times (7000 BC – the date given to the thirteenth plateau), and perhaps even further, and which would thus have been in direct contact with hunter-gatherers. Jacobs suggests that it is in such States that seeds were first gathered, hybridised and finally planted, initially in the soil around the city, and then expanding into the countryside. To explain (and exorcise) the prevalence of the 'agriculture first' dogma, Jacobs draws an analogy with the technologies of electricity (46). Electricity was invented in cities, yet it is primarily in rural areas that we find the massive installations needed for generating and transmitting electricity: dams, power plants, grids. If human memory did not extend back to a time when the world had cities but no electricity, the archaeological evidence could be interpreted to imply that, initially, there were rural people with no electricity; who then developed dams and power plants, eventually producing a large enough surplus of electricity to make cities possible. We are doing something similar when we claim that an agricultural surplus made the State possible, but the error is clear: we turn the *results* of State activity into a *precondition* for the State.

The French historian Fernand Braudel, in his *Civilization and Capitalism*, took up a modified version of Jacobs's thesis, although he was writing in a different context, analysing the relation between the urban and the rural in fifteenth- to eighteenth-century Europe. Braudel likewise contested the dogma that the countryside 'necessarily preceded the town in time' but argued, not that cities preceded the countryside, but that the two were *reciprocally determined*. 'Jane Jacobs, in a persuasive book, argues that the town appears at least simultaneously with rural settlement, if not before it . . . Town and countryside obeyed the rule of "reciprocity of perspectives": mutual creation, mutual domination, mutual exploitation according to *the unchanging rules of co-existence*' (Braudel 1992: 484, 486; cf. Smith et al. 2014: 1532: 'Agriculture and urbanism . . . developed in tandem'). Although Jacobs and Braudel focused their analyses on cities, Deleuze and Guattari (who distinguish between the State and Cities) will adopt a variant of Braudel's thesis with regard to the State: not that the State preceded agriculture, but that agriculture and the State were *co-determined*. 'It is the State that creates agriculture, animal raising, and metallurgy; it does so first on its own soil, then imposes them on the surrounding world . . . It is not the State that presupposes a mode of production, it is the State that makes production a "mode." The last reasons for presuming a progressive development are invalidated' (1987: 429). If the State does not appropriate an already-existing surplus, it is because *the State itself creates the conditions that make a surplus possible*.

Deleuze draws on this second argument when he assesses Friedrich Engels' famous 1884 book on the *Origin of the Family, Private Property, and the State* (Engels 1972). In addition to an agricultural surplus, Engels appealed to several additional sets of factors to explain the origin of the State: *exogenous* factors such as the need to organise wars; *endogenous* factors such as the rise of private property and money; and *specific* factors, such as the emergence of 'public functions'

(Deleuze 1979a; cf. Deleuze and Guattari 1987: 427). But Deleuze shows how each of these factors, far from explaining the emergence of the State, in fact *presuppose* an already-existing State. States can and often do appropriate a war-machine for themselves, but such an appropriation presupposes that the State already exists. Similarly, no one has ever indicated a mechanism through which one could move from a communal tribal property to private property, as if one day, some exceptional person decided to proclaim, 'This is mine'. On the contrary, archaeology has been able to provide a precise mechanism, assignable if variable, showing how private property was constituted out of a system of imperial public property through *freed slaves* – but this means that the privatisation of property could become a characteristic of the State only if the public property of the archaic State were *already given* (Deleuze and Guattari 1987: 449, 451). The same is true for the origin of money, which was created not to promote commerce, but for the purposes of taxation, which likewise presupposed an already-existing State. Finally, public functions also presuppose a State: irrigation, for instance, was an agricultural problem that went beyond the capacities of most agricultural communities.

These analyses all point to the same antinomy: on the one hand, the State could not have emerged from the soil of primitive societies, since they are directed *against* the State; on the other hand, the factors typically put forward to explain the emergence of the State (not only a prior agricultural surplus, but also the military, private property, money, public works and so on) in fact presuppose *an already-existing State*. Every explanation of the origin of the State is tautological, presuming what it seeks to explain. How then can we explain the appearance of the State, if it was not the result of a progressive evolutionary process, and if it 'leads back to no *distinct* assignable cause' (Deleuze and Guattari 1987: 427)? Deleuze and Guattari draw the only possible conclusion: the State appeared in the world fully formed and fully armed, as if it were born an adult, 'a master stroke executed all at once' [*coup de maître en une fois*] (Deleuze and Guattari 2009: 217; cf. 1987: 427). But what, then, does it mean to say that the State appeared in the world 'fully formed'?

## The State is Self-Presupposing (The Apparatus of Capture)

Deleuze and Guattari's second thesis is a correlate of the first: if the State does not evolve from other social formations, it is because *the State creates the conditions of its own existence* (Deleuze and Guattari 1987: 446). Although *Anti-Oedipus* had proposed the term 'overcoding' to describe the basic mechanism of the archaic State, the concept of 'capture' introduced in *A Thousand Plateaus* is meant to provide a more detailed account of the way in which overcoding works. The archaic State, as a self-conditioning entity, is a transcendent *apparatus of capture* that incorporates everything into its form of interiority through three primary abstractions – rent, labour and money – which are a variant of the 'trinity formula' analysed by Marx in the third volume of *Capital* (1981: 953), and operate through two interrelated operations: *direct comparison* in the form of abstract quantities, and *monopolistic appropriation* in the form of stock. A brief review of these abstractions is enough to bring to light their self-presupposing nature.

*Ground Rent.* Rent is a mechanism of capture that allows land, or the 'earth', to be incorporated into the State apparatus. But if we understand the earth as an abstract general space – the *geo* in geometry – we must say that this abstract space was *created* by the apparatus of capture. 'Before' the earth, the land was occupied or territorialised without being measured or divided: there were only the shifting territories of primitive societies, or the smooth spaces occupied by nomadic societies. But the State can claim that these territories and their occupation *already* coexist in a general and abstract space, which is a space that belongs to the despot. Moreover, the constitution of the earth (*geo*) is coexistent with its measurement and striation (*metron*). Every year in Egypt, after the Nile floods, land surveyors or 'rope-stretchers' (*hardenonaptai*) would re-striate the land; the Greeks called them, precisely, the 'measurers of the earth' (*geo-meters*) (Serres 1993). The striation of the earth – its division and portioning out in plots – was the condition for the extraction of rent and tribute, since rent requires a *direct and quantitative comparison* of yields to be drawn between qualitatively different lands. States are often seen as territories centred on the palace-temple complex of a capital city, but more properly one must say that the State 'deterritorialises' the surrounding territories and subordinates them to an imperial centre of convergence located outside and beyond them (the despot as the owner of all the earth).

*Labour.* Similarly, human activity is appropriated by the State in the form of surplus labour, which is stockpiled in large-scale public works projects (pyramids, irrigation projects). The State thus implies a specific mode of human activity that does not exist elsewhere: labour. In primitive societies, strictly speaking, people do not 'labour', even if their activities are highly constrained and regulated. Deleuze and Guattari call this non-labour mode of activity *free action*, which is in continuous variation: one passes from speech to action, from a given action to another, from action to song, from song to speech, from speech to enterprise, 'all in a strange chromaticism with intense but rare peak moments, or moments of effort that the outside observer can only "translate" in terms of work' (Deleuze and Guattari 1987: 491; cf. Guéroult 1934: 119ff.). For labour to exist, there must be a *capture* of such human activity by the State apparatus: it is only in the State that activity comes to be compared, linked and subordinated to a common, homogeneous and abstract quantity called 'labour'. The Egyptian pyramids were not constructed by slaves but by conscripted Egyptian labour, and as such they constitute a form of stockpiled activity. There is no labour outside of the State apparatus, and human activity is transformed into labour only in relation to the State.

*Money*. Finally, just as labour does not exist outside the State, neither does money. Money was not introduced in order to serve the needs of commerce, as if there were first an autonomous domain of 'markets', into which money was introduced to facilitate exchange (Graeber 2011: 44–5). Rather, the converse is the case: money was created by the State to make taxation possible. Money, as an abstract equivalent or unit of account, is an instrument of measure (*metron*) that makes possible a direct comparison between goods and services, which the State can then appropriate in the form of taxes or tribute (Will 1955; Foucault 2013: 133–48; Deleuze and Guattari 2009: 197; 1987: 442–3). For this reason, it is money that creates markets, and not vice versa: the 'economy' presupposes the State. As Litaker observes, money striates space-time through the emergence of markets, which are spaces of commercial exchange that determine the times of production, circulation and consumption (2014: 121).

In short, ground rent, labour and money, in the archaic State, are abstract mechanisms of capture and stockpiling – ground rent captures the land, labour captures human activity, and money captures economic transactions – and each of these mechanisms converges on the person of the despot, who is at once 'the eminent landowner, the entrepreneur of large-scale projects, and the master of taxes and prices' (Deleuze and Guattari 1987: 444). From the viewpoint of Deleuze and Guattari's critique of social evolution, the State's apparatus of capture has several distinct characteristics.

First, and most importantly, *the apparatus of capture creates what it captures*. The earth, labour and money are the conditions that make possible rent, surplus labour (profit) and taxes, but these conditions are themselves created by the State. This is why 'capture' does not simply mean an 'appropriation' of what already exists; both in fact and in principle, the State is only able to capture what it itself creates, or at least what it contributes to creating (446). The State plays the role of a foundation, but it cannot play this role if it captures what already exists: if something exists before the State, it can exist without the State. For the State to be foundational, *the State must be self-presupposing* (427), and it is the self-presupposing nature of the State that grounds its monopoly power, its triple possession *in principle* of the totality of the earth, the totality of labour and the totality of money. The monopoly power of the State can be expressed philosophically in several ways: in the language of sufficient reason, *the State is its own ground*; in the language of causality, the State is *causa sui* (Lampert 2011: 157); in the language of Kant, *the State produces its own conditions of possibility* (and thus is in itself unconditioned).

Second, the apparatus of capture is primarily a *semiological* process of abstraction (Deleuze and Guattari 1987: 445). For Deleuze, every social formation is both a physical system (a manner of occupying space and time) and a semiological system (a 'regime of signs'). In the codes of primitive societies, these signs were inscribed directly on the body in the form of markings (tattoos, circumcisions, incisions, scars, mutilations, and so on) that indicated one's position in the social formation – an entire system of 'mnemotechnics' (Deleuze and Guattari 2009: 144–5). If the ancient despotic State was able to overcode these existing codes, it was because it operated with an abstract and externalised semiotics based on numeracy, literacy and money: the development of geometry and arithmetic, the invention of phonetic writing, the issuing of currency. Money is an abstraction that functions as an abstract equivalent for all goods and services. Geometry treats the earth as an abstract space in which all places are equivalent. Labour allows for a quantitative and abstract comparison of all human activities. Taken together, these three heads of the apparatus of capture create an abstract locus of comparison in which land, goods, services, transactions and human activities are equalised, homogenised, compared, appropriated and stockpiled – all in a single process. In other words, the State operates by abstraction and *is itself an abstraction* (Sibertin-Blanc 2013: 50).

Third, the self-presupposing and abstract nature of the State entails a particular type of *violence*, one that is itself posited as preestablished and preaccomplished, even if it must be reactivated every day. It is often said that the State has a monopoly on legitimate violence – violence against 'criminals', violence against those who capture something they have no 'right' to capture – which the State self-regulates through the institution of law. But this juridical coding of violence *within* the State takes place within the structural violence *of* the State itself, whose apparatus of capture simultaneously constitutes and presupposes a *right* to capture (Deleuze and Guattari 1987: 448). The State, as self-presupposing, is itself a kind of originary or primary violence that is always-already present, even if it never actually 'took place' (see Derrida 2002). As such, it is first and foremost in *myth* that the primary violence of the State finds expression, retrojected in an original violence against chaos that, at the limit, never actually occurred, even if it is omnipresent in every mechanism of the State. Hence the appeal to Dumézil's classic analyses of the two poles of sovereignty found in Indo-European myths: the jurist-kings who operate through law and a respect for obligations, but also the terrifying magico-religious sovereign who operates through a *magical capture* that 'binds without combat' (Dumézil 1988: 152; Deleuze and Guattari 1987: 424–5). The originary and self-presupposing violence of the State makes resistance almost impossible, and it is what gives the State its ultimate power (*puissance*): territorial power (monopoly of the earth), economic power (monopoly of labour), monetary power (monopoly of currency) and, ultimately, political power (monopoly of violence).

## There Has Only Ever Been One State

If the various social formations analysed in *Capitalism and Schizophrenia* do not represent the *evolutionary stages* of social development, neither can they simply be identified as the *ideal types* of a comparative sociology, despite appearances, since each type functions in a different manner (Sibertin-Blanc 2010: 114). The concept of the 'primitive', for instance, can be seen as a type whose unity is the *unity of reason*, theoretically subsuming under a single concept a plurality of heterogeneous societies. By contrast, the capitalist type has a unity that is not only theoretical but also historical: it is a *singular universal*, in the sense that it is the result of a historically contingent process that has resulted in the universalisation of its singularity (Deleuze and Guattari 2009: 140n.). But Deleuze and Guattari ascribe to the State a unity of a completely different nature: a *real unity* that, whether actualised or virtual, is omnipresent throughout the entire social field, not only in archaic States or modern nation-States, but even in primitive societies 'without a State'.

This brings us to a third challenging thesis proposed by Deleuze and Guattari: *there has only ever been but one State*. This thesis is repeated throughout *Anti-Oedipus* (2009: 214, 220, 261) and taken up again in *A Thousand Plateaus*, and initially seems rather untenable. To understand the thesis, we must again distinguish between its *de facto* and *de jure* aspects. Deleuze and Guattari readily admit that *in fact* there is an extraordinary plurality and diversity of existing States, and that modern nation-States, for instance, are very different from the archaic imperial State. But these *de facto* differences between concrete States find their *de jure* ground in the ideality of a single State (pluralism = monism), which Deleuze and Guattari call the *Urstaat* (*ur-* [proto] + *staat* [state]). The *Urstaat* does not refer to a supposed first state, but rather functions as an Idea (in the Deleuzian sense) that is present, throughout the social field, as a virtuality or problem. For Deleuze and Guattari, 'the general theory of society is a generalized theory of flows' or fluxes (262), and the function of social formations is to *code* these fluxes. The Idea of the *Urstaat*, in turn, lies at the opposite pole to the Idea of a pure flux (schizophrenia): it is *the Idea of a completely captured and coded flow*, which is 'the eternal model of everything the State wants to be and desires' (217). As such, however, the pure Idea of the *Urstaat* as such has never been fulfilled in any actually existing State, including the archaic imperial state, which simply managed to actualise the *Urstaat* in its 'purest conditions' (198). In Deleuze's terminology, the Urstaat is an *immemorial* Idea, that is, *a past that has never been given as such* (second synthesis). For this reason, the *Urstaat* itself 'appears to be set back at a remove from what it transects and from what it resects, as though it were giving evidence of another dimension, *a cerebral ideality* [in the Platonic sense] that is superimposed on the material evolution of societies, *a regulating idea* [in the Kantian sense] or principle of reflection (terror) that organizes the fluxes into a whole' (219).

But this is also why the *Urstaat* necessarily functions as a principle of *difference*: every existing (*de facto*) State actualises the (*de jure*) Idea, or resolves the problem of capture, in a different manner. There is thus an internal 'becoming' or mutation of the State-form, but this is a mutation that does *not* constitute a progressive evolution. Rather, the principle of this mutation comes from the same process of capture that defines the archaic State, but functions as its supplementary double: the archaic State cannot overcode and capture without at the same time freeing up a large quantity of decoded flows that escape from it. It cannot create large-scale public works without a flow of independent *labour* escaping from its hierarchised bureaucracy of functionaries, notably in the mines and in metallurgy. It cannot create coinage without flows of *money* escaping, and nourishing or giving birth to other powers (notably in commerce and banking). It cannot create a system of public property without a flow of *private* appropriation growing up beside it, and then starting to slip through its fingers. Finally, it is with the rise of private property that *classes* appear, since the dominant classes are no longer part of the State apparatus, but become distinct determinations that make use of a now-transformed apparatus.

In a multitude of forms, in other words, *the apparatus of capture inevitably gives rise to decoded flows that escape the apparatus of capture* – flows of money, flows of labour, flows of property, flows of population (Deleuze and Guattari 1987: 449; 2009: 223). If the first great movement of deterritorialisation appears in the overcoding performed by the despotic State, the second movement appears in the decoding of the flows that are set in motion by the despotic State's own apparatus of capture. This is the 'paranoid' vector that is inherent in the State-form (Deleuze and Guattari 2009: 193): the State is at once capture and the impossibility of complete capture, since the State can only overcode by decoding (abstraction). The State cannot presuppose itself without *also* presupposing what escapes its form of interiority, namely, *decoded flows*, which are the figure of the 'outside' (*dehors*) of the State, the inverse of its Idea. Just as the State creates what it captures, it creates what *escapes* its apparatus of capture: it is the State's form of interiority (capture) that at the same time creates the State's absolute outside (decoded flows).

It is this situation, internal to the Idea of the *Urstaat*, that gives rise to an incredible diversity of State-forms – 'evolved empires, autonomous cities, feudal systems, monarchies' (Deleuze and Guattari 1987: 459) – all of which will have as their aim the *recoding*, by means of regular or exceptional topical operations, of the products of these decoded flows. Such states have apparatuses of capture quite different from those found in archaic states. Greek city-states, for instance, are phenomena of 'trans-consistency' defined by immanent networks of maritime and commercial circuits, and they mark a new threshold of deterritorialisation in that the flows (of matter, money, labour) that enter and exit the cities (polarisation) must be deterritorialised enough to be captured in the circuits (432). Moreover, as Marx showed, capitalism appeared when the *generalised* decoding of flows set loose by the State reached a threshold of consistency that allowed two of these flows – abstract capital and naked labour – to conjugate in a differential relation. Capitalism would thus require a new 'regime of signs', a new form of abstraction, that would be able to deal with unqualified flows that have no specifiable content, and Deleuze and Guattari argue, famously, that it was only *axiomatics* that could play the role of a new apparatus of capture adequate to the capitalist formation. In all these cases, the thesis that there has only ever been one State has as it correlate the extraordinary plurality of existing States (monism = pluralism).

## The Urstaat Was Active 'Before' its Existence

But this theory of the *Urstaat* raises a complex question regarding the status of the socio-political field of coexistence. If States did not evolve out of more 'primitive' societies, what exactly is the relation between these two types of coeval formation? More precisely, how can Deleuze and Guattari argue that the *Urstaat* is present *throughout* the social field, even in primitive societies, if such societies actively ward off the State? Deleuze and Guattari's response to this question can be summarised in another provocative thesis: *the Urstaat was active 'before' its existence.* If primitive societies ward off the State, they must nonetheless have a 'presentiment' of the State as a limit they are avoiding – a limit they could not reach without self-destructing. The way the *Urstaat* is actualised in historical States is quite different from the way the *Urstaat* pre-exists as a warded off limit in primitive societies. Objectively, Deleuze and Guattari initially explain this phenomenon from a model drawn from physics. If one considers the social field as a field of vectors, one could say that primitive societies are traversed by a centripetal wave that converges on a point $x$ – a point where the wave would cancel itself out and be inverted into a divergent and centrifugal wave, which is a reality of another order (the State) (Deleuze and Guattari 1987: 565 n.14). The point of convergence marks a *potential* or a *threshold of consistency*, and the convergent wave has the double property of both anticipating it and warding it off. The State is thus 'beyond' primitive groups, but 'beyond' does not mean 'after'. The threshold of consistency has always existed, but primitive societies are content to keep that threshold at a distance. We must thus conceptualise the *contemporaneousness* of these two inverse movements on the field of coexistence, 'as if the two waves that seem to us to exclude or succeed each other unfolded *simultaneously* in an "archaeological," micropolitical, micrological, molecular field' (431).

But there is a second issue that comes to the fore here, which is more subjective. Since every exchange of objects requires a way one can *compare* the objects of exchange, no political economist can avoid the question: How should one evaluate the criteria of exchange? Responding to this question requires a theory of *collective evaluations*, or what one might call, in a Kantian vein, 'anticipations of social perception' (Deleuze 1979b). In the Marxist theory of labour value, the way to compare exchanged objects – for instance, an iron axe and a steel axe – is to compare the labour time that is socially necessary for their production, which requires a collective evaluation of both the worker and the entrepreneur using a scientific (or pseudo-scientific) form of quantification. In primitive societies, however, this route is closed off in advance, not because a measure is lacking, but because there is no 'labour time' to be measured. Human activity is in constant variation, and there is nothing that corresponds to labour, much less to labour time.

On this score, Deleuze and Guattari appeal to the nineteenth-century neo-classical theory of *marginalism* which was originally invented to account for the equilibrium of prices within the capitalist regime. If Marx held to the classical theory in which the value of commodities is derived from the quantity of labour required to produce them, marginalists like Stanley Jevons argued that value should instead be analysed in terms of the utility of the 'last' or 'marginal' object (Clarke 1982: 147–50). Business owners know that, beyond a certain limit, the structure of their business will have to change: there are thresholds beyond which an 'assemblage' [*agencement*] cannot maintain its current consistency. For example, how many cows can a dairy farmer purchase without making any changes in his business, such as adding acreage or procuring more equipment? The last cow he could currently buy is the 'marginal object', since if he purchases more, he will have to fundamentally alter the size and structure of his business. More importantly, it is his *anticipation* of the last or marginal cow that determines the price he is willing to pay for the cows he currently needs. If his business can only sustain twenty additional cows, he will not buy fifty, even if their price is substantially discounted. In marginalism, it is the evaluation of the *idea* of the last or 'marginal' object that determines the value of the entire series of real terms.

Though Deleuze and Guattari find marginalism weak as a general economic theory, they find a new field of application for a modified marginalism in non-capitalist formations (Deleuze and Guattari 1987: 437). In primitive societies, they argue, the object of collective evaluation is not labour time, but rather the idea of the last object or marginal object that governs the series of exchanges, and *agriculture is incapable of entering into these serial schemas*. We can thus conceptualise a difference between the 'limit' and the 'threshold': in a collective evaluation, what is anticipated is the *limit* (the penultimate exchange, which allows one to remain in the same assemblage) but what is warded off is the *threshold* (which would force one to change assemblage). 'It is the evaluation of the last as limit that constitutes an anticipation and simultaneously wards off the last as threshold or ultimate (a new *agencement*)' (439). The threshold marks the point where stockpiling would begin, and the temporal succession of territories would be replaced by the spatial coexistence and exploitation of different territories: the apparatus of capture.

In both these analyses – objective and subjective – we can see how the State has a positive status in primitive societies as both a limit and threshold, even if the State does not 'yet' have an actual existence.

## The Field of Coexistence: Types, Powers and Becomings

The principle behind Deleuze and Guattari's critique of traditional (chronological) theories of social evolution can be summarised in a final thesis: '*All history does is to translate a coexistence of becomings into a succession*' (Deleuze and Guattari 1987: 430; cf. Lundy 2012). But what exactly is the nature of the *field of coexistence* presumed by Deleuze and Guattari's socio-political philosophy? We can perhaps distinguish three levels in their analysis of the field of coexistence, which begins with *types*, then evaluates their *powers*, and finally maps out their *becomings*.

At the first level, *Anti-Oedipus* initially presents us with a typology of social formations, and these 'types' can be understood in a Bergsonian manner. In *Matter and Memory*, Bergson created his well-known concepts of 'pure memory' and 'pure perception': although perception and memory are always mixed together in experience (*de facto*), these concepts allowed him to distinguish the differences in nature (*de jure*) between the two lines or 'tendencies' of pure memory and pure perception. The same is true for Deleuze and Guattari's typology of social formations. Although each type *in fact* coexists with the others within a single field of coexistence – in our contemporary situation, States, war-machines and archaic territorialities all coexist within the capitalist axiomatic – each concept is a tool that allows one to demarcate distinctions or differences in kind within the social multiplicity (Bogue 2004: 172–3).

At a second level, however, *A Thousand Plateaus* characterises each of these types in terms of a specific 'machinic process': primitive societies are characterised by mechanisms of anticipation/prevention; States are characterised by apparatuses of capture; nomadic war-machines by the occupation of smooth space; cities by instruments of polarisation; ecumenical organisations by the encompassment of heterogeneous formations; capitalism by decoding/axiomatisation. This is no longer a question of types; rather, each of these processes is a *power* [*puissance*] that indicates a certain *capacity* or *capability* of a social formation. Primitive societies *anticipate* and *ward off*, archaic States *capture*: this is what they 'can do', what they are capable of. In a Spinozistic manner, each of these powers or processes must be grasped *positively* as a determinate quantity of reality (see Sibertin-Blanc 2013: 41–6). One problem with evolutionary schemas is that they tend to view social formations through the prism of the State-form, which leads to the litany of 'societies without' – 'without a State', 'without history' 'without writing'. But this focus on the State-form winds up assigning privation and lack to other formations, severing them from the forms of power that each of them affirms positively. The second level thus takes us from Bergson to Spinoza: beneath the categorial typology of social formations, one finds an ethological map of their constitutive powers, 'a logic of coessential positivities and coexisting affirmations' (Deleuze 1988: 95).

But the third level is uniquely Deleuzo-Guattarian. Far from being governed by a single form of power, every social formation, both in fact and in principle, is composed of a *plurality* of processes that are in 'perpetual interaction' with each other (Deleuze and Guattari 1987: 430), and each process can function at a 'power' other than its own. Anticipation-prevention mechanisms, for instance, 'are at work not only in primitive societies, but are transferred into Cities that ward off the State-form, into the State that wards off capitalism, and into capitalism itself, which wards off and repels its own limits' (437). Similarly, the State is able to capture, not only land, activity and exchange, but also the anticipation-prevention mechanisms themselves, as well as the war-machine and the instruments of polarisation that characterise cities. And the powers 'become' something other when they enter into relations with each other: the power of the war-machine changes nature when it is 'appropriated' by the State, just as the State's apparatus of capture changes nature when it is subordinated to the worldwide capitalist market. This is the sense of the term 'becoming': it is what takes place *between* two multiplicities, changing their nature. What appears in evolutionary theories as a chronological succession is, for Deleuze and Guattari, a phenomenon of transfer or transport between becomings. In each case, one must ask: What is a social formation capable of? What can it tolerate or support? What are the processes that exceed its capacities for reproduction, and put it in question? When does it pass its limit and enter into a new threshold of consistency? How does it *become*?

Thus, we have to say that the term 'field of coexistence' does not simply refer to an external and *de facto* coexistence of social formations in a historical space-time, but more profoundly to an intrinsic and *de jure* coexistence of powers and processes in a non-historical space-time, a continuum in which divergent temporalities coexist. This is what Deleuze calls the 'plane of immanence', a field where all the powers of THE social machine coexist virtually, in constant becoming, enveloped and implicated in each other in 'a topological space and a stratigraphic time' (Lapoujade 2014: 218).

## Notes

1. In *Capitalism and Schizophrenia*, Deleuze and Guattari continued to use terms such as 'primitive' (French, *sauvage*) and 'barbarian', which have largely been abandoned by anthropologists, presumably in order to emphasise more starkly the difference between their approach and that of 'progressivists' such as Morgan.
2. Both the second chapter of *Difference and Repetition*, entitled 'Repetition-for-Itself' (Deleuze 1994: 70–128), and *Cinema II: The Time-Image* (Deleuze 1989) develop in detail Deleuze's metaphysics of time.
3. Deleuze's concept of actualisation is deeply indebted to the work of Raymond Ruyer, most notably his 1946 book *Éléments de Psycho-biologie*. See in particular chapter 6 ('The Problems of Actualization'), and the important section on 'Actualization and Time' (109–14): 'In the bio-psychological order, the real fact is the passage to the actual . . . of a potential which is not itself in time, although it is progressively modified by its own actualizations' (110).

## References

Bogue, R. (2004), 'Apology for Nomadology', *Interventions* 6: 2, pp. 169–79.

Braudel, F. (1992), *Civilization and Capitalism, 15th–18th Century*, Vol. 1: *The Structure of Everyday Life*, trans. S. Reynolds, Berkeley: University of California Press.

Carneiro, R. L. (2003), *Evolutionism in Cultural Anthropology: A Critical History*, Boulder, CO: Westview.

Childe, V. G. (1951), *Man Makes Himself*, New York: New American Library.

Childe, V. G. (2009), *The Prehistory of European Society*, Nottingham: Spokesman.

Clarke, S. (1982), *Marx, Marginalism, and Modern Sociology*, London: Macmillan.

Clastres, P. (1989), *Society Against the State: Essays in Political Anthropology*, trans. R. Hurley with A. Stein, New York: Zone Books.

Clastres, P. (1994), 'Primitive Economy', *The Archaeology of Violence*, trans. J. Herman, New York: Semiotext(e).

Deleuze, G. (1979a), Seminar of 6 November, *On State Apparatuses and War-Machines*, <http://www2.univ-paris8.fr/deleuze> (last accessed 17 July 2017).

Deleuze, G. (1979b), Seminar of 27 November, *On State Apparatuses and War-Machines*, <http://www2.univ-paris8.fr/deleuze> (last accessed 17 July 2017).

Deleuze, G. (1988), *Spinoza: Practical Philosophy*, trans. R. Hurley, San Francisco: City Lights Books.
Deleuze, G. (1989), *Cinema II: The Time-Image*, trans. H. Tomlinson and B. Habberjam, Minneapolis: University of Minnesota Press.
Deleuze, G. (1994), *Difference and Repetition*, trans. P. Patton. New York: Columbia University Press.
Deleuze, G. and F. Guattari (1987), *A Thousand Plateaus*, trans. B. Massumi, Minneapolis: University of Minnesota Press.
Deleuze, G. and F. Guattari (2009), *Anti-Oedipus*, trans. R. Hurley, M. Seem and H. R. Lane, New York: Penguin Classics.
Derrida, J. (2002), 'Force of Law: The "Mystical" Foundation of Authority', *Acts of Religion*, ed. G. Anidjar, New York: Routledge, pp. 230–98.
Dumézil, G. (1988), *Mitra-Varuna*, trans. D. Coltman, New York: Zone Books.
Engels, F. (1972), *The Origin of the Family, Private Property, and the State*, New York: International Publishers.
Foucault, M. (2013), *Lectures on the Will to Know: Lectures at the Collège de France, 1970–1971*, ed. D. Defert, trans. G. Burchell, New York: Palgrave.
Graeber, D. (2011), *Debt: The First 5,000 Years*, Brooklyn: Melville House.
Guéroult, M. (1934), *Dynamique et métaphysique leibniziennes*, Paris: Les Belles Lettres.
Jacobs, J. (1970), *The Economy of Cities*, New York: Vintage.
Lampert, J. (2011), *Deleuze and Guattari's Philosophy of History*, London: Bloomsbury.
Lapoujade, D. (2014), *Deleuze, les mouvements aberrants*, Paris: Minuit.
Litaker, J. (2014), 'Capitalism and Social Agency', PhD dissertation, West Lafayette: Purdue University.
Lull, V. and R. Mico (2011), *Archaeology of the Origin of the State: The Theories*, Oxford: Oxford University Press.
Lundy, C. (2012), *History and Becoming: Deleuze's Philosophy of Creativity*, Edinburgh: Edinburgh University Press.
Marx, K. (1965), *Pre-capitalist Economic Formations*, trans. J. Cohen, New York: International Publishers.
Marx, K. (1981), *Capital: Volume 3*, trans. D. Fernbach, London: Penguin.
Mauss, M. (1954), *The Gift: Forms and Functions of Exchange in Archaic Societies*, trans. I. Cunnison, London: Cohen and West.
Morgan, L. H. (1877), *Ancient Society; Or: Researches in the Lines of Human Progress from Savagery through Barbarism to Civilization*, London: Macmillan.
Ruyer, R. (1946), *Élements de psycho-biologie*, Paris: Presses universitaires de France.
Sahlins, M. (1972), *Stone Age Economics*, Chicago and New York: Aldine Atherton.
Serres, M. (1993), *Les origines de la géométrie*, Paris: Flammarion.
Sibertin-Blanc, G. (2010), *Deleuze et l'Anti-Oedipe: la production du désir*, Paris: Presses universitaires de France.

Sibertin-Blanc, G. (2013), *Politique et État chez Deleuze et Guattari. Essai sur le matérialisme historico-machinique*, Paris: Presses universitaires de France.

Smith, M. E., J. A. Ur and G. Feinman (2014), 'Jane Jacobs' "Cities First" Model and Archaeological Reality', *International Journal of Urban and Regional Research* 38: 4, pp. 1525–35.

Ulman, G. L. (1978), *The Science of Society: Toward an Understanding of the Life and Work of Karl August Wittfogel*, The Hague: Mouton de Gruyter.

Will, E. (1955), 'Refléctions et hypothèses sur les origines du monnayage', *Revue numismatique* 17, pp. 3–24.

Wittfogel, K. A. (1957), *Oriental Despotism: A Comparative Study of Total Power*, New Haven: Yale University Press.

Chapter 8

# Epigenesis and the Outside

*Claire Colebrook*

There are two ways, at least, that one might think about the relationship between philosophy and life. The first, to follow Foucault, would be that at a certain point in the history of thought philosophy no longer simply has life as one of its possible topics, but something like an ethics of life becomes possible: what we ought to do, who we are and what would count as normative depends upon normal life (Foucault 2002: 139). The second theory, and perhaps one that is becoming the cultural dominant, is that a correct account of life attained by the life sciences should then prompt philosophy to rethink its notions of selfhood, ethics, mind and matter. One might think of the first as critical biopolitics (as adopted by Foucault and to some extent Giorgio Agamben [1998]), and the second as affirmative biopolitics (most explicitly articulated by Roberto Esposito [2008], but more widely and implicitly endorsed by new vitalisms, affect theory and some readings of Deleuze). In this essay I want to refuse the exclusive disjunction that suggests that either life is bracketed and left out of play or that life provides philosophy with its proper ground. Neither correlationism nor a philosophy of biology that takes its cue from knowledge of life; one might think of the two terms – philosophy and life – as both really distinct and yet never fully separated, neither capable of providing the ground for the other. If this were so, then rather than seeing the turn to life as philosophy finally stepping outside itself, and rather than seeing the suspension or bracketing of life as philosophy finally achieving anti-foundationalism, one might argue for an epigenesis that 'goes all the way down'. Here, the philosophy and biology that studies life would be in constant transformation, and the life studied by philosophy and biology would already be philosophical, with each aspect of 'life' already open to the non-living, already captured by ideas and forms.

The theory of life that, for Foucault, enables biopolitics, normalisation and ongoing failure to consider the ways in which 'life' comes to be generated as an object of knowledge, is a life of bounded organisms, where capacities – such as thinking or speaking – would be explicable according to a broader force that operates to sustain and maintain itself. How might this criticism of biopolitics be transformed if instead of being a substrate that operates to further and maintain individuals, life were more like Foucault's own description of power? What if life were not bodies subtended by some driving life force that *then* come into relation for self-maintenance, but a complicated milieu of encounters, relations, forces, interactions and dependencies where a living being might be not only more like a population, but also not so much a being as a relatively stable point in a dispersed field of forces with no intentionality, and certainly not anything like a life drive or genetic selfishness? One might say that the problem with biopolitics is not so much that life becomes a normalising and legislating ground for political managerialism, but that 'life' has been manufactured as a political figure – as a self-driving and self-maintaining force. Perhaps a more accurate – a more scientific – account of life would attend to life's complexity and would therefore disable all the managerial and neoliberal imperatives that allow politics increasingly to become nothing more than a practice of maximising life. In short, is there still a legitimacy to the criticism of *Lebensphilosophie* that runs from Husserl and Heidegger, through to Foucault and Derrida, that insists that no theory of life can explain *knowledge* of life? Or, would a more sophisticated account of life – one that went beyond bounded organisms, one more oriented to developments irreducible to organic efficiency – be able to explain the complex events of philosophy, art and science? I want to suggest that it is only if life is considered in its normalising and organicist sense that biopolitics paves the way for a neoliberalism that sees market forces as enabling a survival of the fittest. If life were neither the self-maintaining body nor a substrate of forces then one might arrive at a radical or general epigenesis. Epigenesis would not only be what 'we' might observe about life, but would provide a way of thinking beyond life: if, as epigenesists propose, individuals do not unfold from a blueprint but are forms that transform in relation to other forms, one would have to abandon the conception of life 'as such'.

There's quite a difference between saying that humans are rational animals, where rationality is a capacity for *thinking as such* regardless of who or what undertakes the act of thinking, and the modern sense of an evolved rationality where the tendencies of thinking might be explained by reference to a broader logic of life, and where rationality would be a specific and species-grounded tendency explicable in terms of human historical development (Foucault 2002: 140–1). Should one think of reason and logic as purely formal, real and transcendental powers that allow for something like a pure subject who happens to be embodied but whose capacity for *thinking* operates regardless of the singular historical individual? Or, should one think of such systems as having emerged gradually from tendencies evidenced in all life, where reason would be yet one more survival mechanism? Here, the formal systems of mathematics would have emerged from practical and bodily habits, eventually forming technologies of calculation, in turn generating systems of more complex modelling and abstraction, but would ultimately be explicable by a theory of purposive life. That is, even if mathematics or logic today might be pure and abstract, their *origin* would lie in human life's embodied, world-oriented and purposive orientation. (Such an account is given by Jürgen Habermas in *Knowledge and Human Interests* [1971]. All knowledge is grounded in the lifeworld and should *always* be held to account according to the ongoing reflective interests of the community. Or, as Bruno Latour has argued: however abstract scientific problems may be, they are nevertheless prompted and enabled by matters of concern [Latour 2004].) It is possible to claim both that purely formal systems have their origin in life, *and* that once constituted they have a consistency and reality that is irreducible to their origin (Tragesser 1984). One might be Deleuzian and refer to this as higher deterritorialisation, insisting both on the corporeal and embodied emergence of systems *and* their capacity to achieve a new stratum of autonomous operation. One might be able to insist *both* on the need for philosophy to take account of life *and* to insist that life provides nothing like a ground.

The possible difference between the two inflections is nevertheless profound. If one stresses the embodied, emergent and distributed nature of reason then it becomes possible to trace and judge the ways in which thinking turns against the life from which it emerged. Think, here, of everything from Marxist conceptions of ideology to popular evolutionary psychology, and all the ways in which experts – blessed with a theory of life – can explain why 'we' so often fail to act in our own interests. If, however, one imagines that (whatever its origins) thought has a capacity to act *as if* it were pure and disembodied one would stress a certain futural and utopian potentiality. That is, rather than argue that because systems emerged from life they should always be adjudicated according to their capacity to further life, one might think of life as a tendency to de-realise, de-nature, and depart from itself. Epigenesis would not simply refer to the ways in which the very essence and interior of the living being becomes something different in relation to a milieu; it would not only require the abandonment of the very ipseity of the individual, generating a new notion of form as a potentiality to differ, but would intensify the power of forces to operate in a rogue, anarchic and lifeless manner with no ground, no inside and no environment. In short, one could *either* see the task of thinking as one of drawing thought back to the life from which it emerged, destroying the illusion of Cartesian man by seeing thought as bound up with practices and modes of existence, and therefore regarding the pure subject as a pernicious notion that has allowed a highly normative conception of (autonomous and able-bodied) individualism to occlude the richness and complexity of the mind in life. *Or*, one could stress deterritorialisation, and the capacity for systems to operate in a different register. One could see deterritorialisation as something like *techne*, or the capacity for bodies to take on systemic and repeatable practices that become autonomous. One could then judge deterritorialisation according to the degree to which it maximised life, *or* one could imagine intensifying deterritorialisation beyond life (or at least beyond organic, individual, purposive life).

Just how one reads the concept of deterritorialisation itself depends both on a theory of life and on how one understands the relationship between life and theory. One possibility, and one that I will argue for here, is that deterritorialised systems should be understood according to a general epigenesis, where stratifications such as language, cities, fashion, war-craft and music are not organic wholes capable of being analysed according to their own internal criteria or coherence but are ongoing dynamic formations in continual response to other formations or assemblages. If this were so then all parts might be considered as living systems; 'life' would include technologies, not because one had expanded the concept of life to included non-organisms, but because what was once thought of as an organism was already made possible by formative and mutually constitutive relations to forces that in no way form organic wholes (Deleuze and Guattari 1987: 41). What were once thought of as living systems would always be in touch with multiple structures once relegated to the realm of the non-living. Another possibility would be to read deterritorialisation as generating a radical break, rupture or event, such that one might think of a force at odds with life, and perhaps at odds with systems or symbiosis in general. What would it mean to posit an outside to life, an other-than-life or non-life that was not simply the organic death that is bound up with the ongoing flux of life, and not simply life's own capacity to generate new and hitherto unimagined relations? 'Absolute deterritorialisation' might be thought of as this tendency within life for movements of separation (constant and ongoing) to be released from organisms and from symbiosis (Deleuze and Guattari 1987: 6). What if philosophy, or cinema, mathematics or art – despite all Deleuze and Guattari's accounts of their emergence – were to produce relations of complexity and intensity that could no longer be regarded as living, *if* life is to be understood as some form of self-maintenance (whether of the gene, the individual or the type). Far more specifically, one might read Deleuze's (and Deleuze and Guattari's) references to genesis as a way of thinking about the relation between life and philosophy. Are all the plateaus that describe refrains, art, insect life, becoming-animal and molecular becomings gesturing toward an ethics of life, where the inhuman genesis of what we take to be so specially human should draw us back to pre-individual forces, *or* are these descriptions of life ways of thinking about intense individuation that creates something like a beyond or outside of life?

I would suggest that readings of Deleuze, and broader claims for the importance of thinking about philosophy in relation to life, assume this problem: it seems that one either stresses Deleuze's vital, embodied materialism, *or* one stresses a power of thought liberated from life. One must either see life as a criterion for judging and understanding systems, or one places life out of play and stresses the capacity for formalism and abstraction. If, however, one regards life as already other than itself, as a deterritorialising or technical power, then this problem of life and non-life, of vital versus abstract systems, needs to be displaced. Rather than draw systems back to life – even in the manner of Bernard Stiegler's epiphylogenesis that regards individuals as having their proper being in a shared, dynamic and collective archive (Stiegler 1998) – and rather than fetishising the break with life (as, say, with Alain Badiou's subject [Badiou 2009]) – life would be always already other than itself. Practices such as philosophy could neither take their cue from biology (as though life were simply an object to be studied), nor could philosophy bracket life and attend to pure forms or an Absolute. A general epigenesis would entail thinking of all thought practices (including philosophy) as at once formed in relation to encounters with an outside *and* as retaining, from each of these encounters, a certain inscribed or sustained tendency.

A quick glance at *A Thousand Plateaus* (insofar as a quick glance is possible) would seem to suggest that the notion of language or signification generating a radical break with life ('the despotism of the signifier') is precisely what is rejected by various concepts in Deleuze and Guattari's work, including the notion of multiple regimes of signs, of stratification and becoming-animal. To a *certain extent* such a reading of Deleuze and

Guattari as offering a theory of life that will undo the dominance of language is fully justified. The proper direction of thought would seem to be away from formalisation and subject formation toward life. If there is a semiosis beyond language in the narrow sense and beyond 'the' signifier, then it is illegitimate to fetishise writing or inscription as a radical break (Deleuze and Guattari 1987: 65). It would also be illegitimate to see life as that which must be bracketed or suspended or put out of play in philosophical questions of who we are, what we might become and how we *ought* to be. One way of reading the significance of Deleuze and Guattari's thought would be not only to see it as a philosophical revolution that takes life seriously, no longer beginning with thought that proceeds from a bounded self, and no longer isolating thought from life processes, but also as transforming the life that would provide a normative ground for how one does philosophy.

Rather than see artworks or poems – as the French avant-garde tradition had done – as purposeless, detached, set apart from and radically at odds with life (where life is connectedness, relation, self-maintenance and symbiotic), Deleuze and Guattari tie art to the animal, and even suggest that something like the refrain (a pulsional rhythm) is life itself: 'This is because one does not think without becoming something else, something that does not think – an animal, a molecule, a particle – and that comes back to thought and revives it' (Deleuze and Guattari 1994: 42). If *life* is characterised by ongoing transformation in relation to other forces that are, in turn, nothing other than responses, then it follows not only that one would read all domains of life as bound up with multiple living systems, but also that philosophy ought to transform itself beyond the detachment of logic and language. This, in turn, would yield something like an ethics of *life* that would be at odds with a dominant tradition in late twentieth-century French philosophy of an ethics of the subject.

Rather than follow a line that runs from Kant through to Derrida and Foucault, where a capacity to *think* or generate concepts would open a realm of freedom that would liberate itself from life (and from the human-all-too-human needs and interests that tie actions to bodies and identities), Deleuze seems to offer a way beyond the exclusive disjunction of either an ethics of life or an ethics of the subject. I would suggest that he does so not by drawing the subject back to life, but by dispersing the subject in life. To take just one possibility: 'becoming-molecular' might be understood as a destruction of the world: one would neither see life as the ground and criterion of action, nor regard human subjects with their formalised, inscribed and archived systems as the horizon or lifeworld through which action might be legitimated. *Philosophy*, then, might itself be an ethics insofar as it would be the creation of concepts that could not be reduced to what might serve and maintain bodies and persons: what might it be to think at a molecular level, where life is not bodily or organic life, but *perceptive life*? 'To live' would be to be affected, but in a manner that would *precede* relatively stable affects and emotions. Affect would *not* be feeling but the perceptive/responsive modification from which feelings would be composed. Every molecule is what it is only in its singular encounters with other molecules. Epigenesis is not only the transformation of ongoing individual forms in relation to a milieu, but a general and contaminated becoming.

Deleuze's ethics of becoming, then, is neither an ethics of creativity (where one thinks of creation as an act, where one brings oneself into being), nor an ethics of life, if life is understood as something that might be maximised or intensified. This is not simply because his account of life yields a different mode of philosophy, but rather because a certain way of thinking about life problematises the relation and possible non-relation between life and philosophy, between what is (being) and the thought of what is. Even though Deleuze and Guattari (in *What is Philosophy?*) grant philosophy a distinct autonomy in its creation of concepts, and even though this same volume seems to place art and sensations closer to life, their account of life is one that already harbours a tendency toward composition and ideality. It is as though even a bird and an array of leaves cannot be thought of as animal and world, but already display a potentiality toward ideality, not only pure sensations but *then* a philosophy that in turn creates the concept of 'percepts'. Even the concept of a life form is not the grasping of a type but an exercise in perception of relations: 'The concept of a bird is found not in its genus or species but in the composition of its postures, colors, and songs: something indiscernible that is not so much synesthetic as syneidetic. A concept is a heterogenesis – that is to say, an ordering of its components by zones of neighborhood. It is ordinal, an intension present in all the features that make it up' (Deleuze and Guattari 1994: 20). When they go on to talk about absolute survey, a term taken from Raymond Ruyer, they draw attention to a conception of form that is neither a genetic blueprint, nor information, nor a phenotype but a diagram of relations, an orientation – the distinct ways in which components become in relation. Philosophy's concept – what philosophy is and does – amounts to a perception of life oriented to components that are what they are only in an ideal relation: 'the concept is defined by *the inseparability of a finite number of heterogeneous components traversed by a point of absolute survey at infinite speed*' (21).

The philosophical concept seems to be defined in relation to what life is – heterogeneous components – and yet life appears to be already proto-philosophical: 'traversed by a point of absolute survey'. One way to think about post-phenomenological philosophy in late twentieth-century France is to see some of the major works as responding to the problem of life and genesis. Husserl's work charts a path between pure logical abstraction (or formalism) and natural emergence (psychologism and historicism). In their early work both Derrida and Foucault problematise the Husserlian notion of a genesis of thought in life, with Derrida insisting on the radical and anarchic break of concepts, and with Foucault charting all the ways in which 'life' is bound up with a normalising biopolitics of 'man'. For both Foucault and Derrida psychologism, naturalism and historicism – or any approach that assumes a plane of life from which structures emerge – domesticate the rogue power of writing (Derrida) or literature (Foucault). Derrida's early work on genesis and life is insistent that there is a rogue inscriptive power that enables any thought of life, and that any attempt – such as Husserl's – to account for the emergence of formal structures from life will always come up against an aporia of presupposing a purity of life that can only be given ex post facto (Derrida 2003: 40–1). Against that Derridean opposition between inscription/writing and life, Deleuze and Guattari see life itself as inscriptive: *not* a code, *not* genetic information, but always retentions and repetitions of differences that result from interactions:

> The recordings and transmissions that have come from the internal codes, from the outside world, from one region to another of the organism, all intersect, following the endlessly ramified paths of the great disjunctive synthesis. If this constitutes a system of writing, it is a writing inscribed on the very surface of the Real: a strangely polyvocal kind of writing, never a biunivocalized, linearized one; a transcursive system of writing, never a discursive one; a writing that constitutes the entire domain of the 'real inorganization' of the passive syntheses, where we would search in vain for something that might be labeled the Signifier – writing that ceaselessly composes and decomposes the chains into signs that have nothing that impels them to become signifying. (Deleuze and Guattari 1983: 39)

In response to Foucault's insistence that it is *literature* that harbours the possibility for thinking about relations beyond the normalising ground of man (because writing would no longer be communicational or functional, no longer driven by life), Deleuze insists that life can generate its own ungrounded dimensions: 'The forces within man enter into a relation with forces from the outside, those of silicon which supersedes carbon, or genetic components which supersede the organism, or agrammaticalities which supersede the signifier. In each case we must study the operations of the superfold, of which the "double helix" is the best known example' (1988: 131–2). In relation to the thought of their own time, Deleuze and Guattari had already refused the problem of *either* naturalism *or* formalism. Yet today this exclusive disjunction has apparently intensified, with new forms of realism, affect theory, emergence theory and naturalised phenomenology occurring alongside AI attempts to free thought or intellect from embodiment, as well as ongoing practices of logic and calculus that operate regardless of (human) bodies. Either one insists that formal systems, thought, language, and mind in general emerge from life and therefore ought to be considered ecologically, *or* one considers thinking to be a power capable of formalisation and abstraction. This would then seem to entail a mode of ethics or politics, either negotiating events according to a broad field of composition and relations (Protevi 2009; DeLanda 2006), or insisting on the power of thought to be sustained beyond life (Bostrom 2014).

It might seem obvious that Deleuze (and especially a Deleuze committed to epigenesis and the force of the double helix) would provide a philosophical foundation for conceptions of mind as emergent, *and* that the plane of life from which structures emerged would offer some sort of ethical criterion. One might think of the ways in which an attention to development (and the role of transformative development through encounters) strengthens the notion that the human mind is *not* an isolated rational capacity but part of a broader milieu. Epigenesis charts a path between a pure constructivism or formalism and a straightforward naturalism. Rather than think of biology as a form of essentialism or determinism, as though life forms followed their own path and then faced selective forces in relation to an outside, every force would be outside itself, already bearing the marks and potentialities of prior encounters. More specifically, a general epigenesis would not only see bodies as inscribed by (and inscribing) surrounding forces, but would regard every force as a 'force from the outside', as at once bound up with a milieu but also as offering a disturbance. The polity would no longer be comprised of isolated decision makers, but would be an ongoing network or assemblage of bodies, human and non-human. Bearing this in mind it would be possible to draw a contrast between a Deleuzian-inspired theory of social/political assemblages informed by the life sciences (where life does not operate as some legislative or normalising foundation) and Foucault's strategy of countering biopolitics. For Foucault, 'life' as the horizon or plane that enables certain knowledge practices is bound up with a politics of normalisation and ultimately neoliberalism. If there is something like 'life' in general there can both be a principle that explains the emergence of complex formations – such as languages and the division of labour – and forms of political practice that are able to deploy expertise to manage populations. Here, Foucault at once extends and reacts critically to Kantian subjectivism. If one were to assume that knowledge of life would give one the power to intuit or legislate the good, knowledge would be operating as a form of *exception* or expertise. One could claim to know better. For both Kant and Foucault, knowledge, as relational, is necessarily intertwined with ongoing practices of verification, reflection and legitimation; to claim that an event in the world is true is to assume that it would continue to count as true for others, and that it coheres and would continue to cohere with already established and ongoing practices of truth. No act of knowledge of the world could step outside and govern these relations. Ethics could not be based on what one claims to know, nor on any theory of life. Any claim or posited law of life proceeds from a subject to whom the world is given; that same subject, however, cannot apply this law of life to what she may or may not decide to do, nor to what others decide to do. The very same capacity that allows one to think of life as a regular, lawful, stable and interconnected causal whole precludes any form of adjudication regarding what one ought to decide, *and* opens a space of subjective freedom.

Despite Quentin Meillassoux's targeting of Kant as the villain of correlationism – an epistemology that would preclude any claims regarding the world as it is in itself – both Meillassoux and Kant insist upon the limits of what might be thought. For both it is not life as it is but *what might be thought* that is philosophy's task. While Kant insists that the *world* would make no sense without ongoing causal coherence, ethics requires that one act *as if* the causal order of the world would not apply to the subject. No matter how coherent, verified or substantiated one's account of life might be, it would always be possible to *think* otherwise. For Meillassoux, the very possibility of scientific statements relies upon a knowledge of life that would hold true, even if there were no consciousness there to experience such a truth, and yet the conclusion he draws from this possibility of thinking life in itself is *not* a form of naturalism but an insistence on the possibility to think of an Absolute. Nothing we know about life is necessary, and while it is the task of the sciences to think about what life might be *as such* and without us, it is philosophy's task to consider the limits of possibility, regardless of what we know of life (Meillassoux 2008).

One might, then, draw a stark contrast today between Deleuzian-inspired accounts of ethics and politics indebted to a theory of life (where life is anything but a biological or natural determinism) and forms of speculation (Meillassoux), anti-biopolitics or accelerationism that stress forces that are other than life. Such a contrast would seem to draw one back to the origins of what has come to be known as continental philosophy and its commitment to considering the sense, meaning or disclosure of life as prior to the positing of life itself. Continental philosophy would mark a break with naivete, the natural attitude or any other foundationalism. To talk about *naturalised* phenomenology, the emergence of mind, or mind in life would rely on some uncritical foundational term (nature, mind, life). As the Husserlian criticisms of Kant insisted, what phenomenology provides is not a theory of the world but of the *origin* of the world: there is no *world* if there is no synthesis, and – more importantly – there is no synthesis without a subject (even if that subject is radically passive) (Fink 1933). The *thought* of life is transcendental, and any theory of life offered by the natural sciences would presuppose a lifeworld and horizon of sense.

Is there a way beyond this disjunction between philosophy as a transcendental horizon that would suspend or bracket life, and a philosophy that takes its cue from life? Would it be possible to transform the problem such that it is no longer a question of whether the life sciences must presuppose some conceptual scheme or whether a theory of life should chasten philosophy's formalism? To what extent does that question of scientific reality and its relation to philosophy already presuppose a certain philosophy of life? I would suggest that as long as one thinks about life as that which is studied by the life sciences, by biology, then philosophy will have to establish some type of relation *to life*. If one were to begin with the conception of bounded and relatively autonomous organisms, then it is possible to think of both subjects for whom there is a world, and a world that is experienced as a stable, causally ordered, rational whole. It would then follow that philosophy as an art of thinking could be relatively autonomous from the content or facts of the world.

Again, for all the ostensible objections against Kantian correlationism and the claims for a new realism, Meillassoux's work is typical of a philosophical tradition that emphasises *what might be thought*, in a mode of radical abstraction from life. In an age of neoliberalism where individual human bodies are regarded as capable of self-optimisation if they are oriented toward the smooth workings of the social *and* living whole of the polity, it is perhaps not surprising that it might seem politically radical to detach a power of thinking from the everyday encounters of bodies and socio-political relations. The notion of a detachable intellect, of superintelligence, of mind in itself, or of the power of pure thinking not only has a Cartesian heritage, but relies upon an ongoing conception of distinct individuals. One might think here of Alain Badiou's conception of the subject; although the subject is clearly *not* the organic individual and is defined as a capacity to break with life, subjectivity is *more* unique, more bounded, more set apart than the physical body (Badiou 2009). Those forms of ethics that defined themselves against life nevertheless rely upon a notion of a bounded subject that is distinct, individuated and capable of thinking beyond bodily interests. Here, Badiou continues a post-phenomenological tradition critical of any attempt to allow *life* to operate as the final horizon for ethics.

If Kant had argued that the world was empirically real – experienced as other than ourselves and therefore capable of being subjected to scientific theoretical claims – this was accompanied by a transcendental ideality that enables one to think (if not know), a capacity that is irreducible to life. Kant's legacy is, therefore, ambivalent. On the one hand there is an emphasis on organicism and bounded bodies, and a world that is experienced in terms of distinct identities, causal relations and ongoing stability (and it is this *correlative* world that comes under attack in the new realisms). On the other hand, and yet connected to the coherence and order of the world, there is a subject whose 'life' is not that of the organic order (Malabou 2016). Organicism at once pertains to the boundedness and coherence of bodies *and* to the ways one thinks of viewing the world – also as a coherent whole, composed of relatively autonomous parts, each with its own logic and life. Despite the interconnectedness of the idea of bounded bodies and the coherence and distinction of subjects as minds, there is a tension that is played out in the problem of the relation between philosophy and life (if one's theory of life is one of organicism). As I have already suggested, despite naming Kant as the first in a long line of correlationists, Meillassoux's insistence upon an absolute that is achieved by a focus on what is *thinkable* at once destroys the notion that a world of coherent identities is the only possible world, and yet intensifies the purity and detachability of thinking. While destroying the connectedness, coherence and organicism of life, a higher coherence and identity is achieved at the level of thinking. By refusing the primacy of life as a philosophical ground one also refuses the embodiment, distribution and living nature of mind.

This brings us back to the apparently stark opposition between theories of life that attend to relations, distribution, interconnectedness, dynamism, symbiosis and the distributed nature of mind, and philosophies of the subject that emphasise the inhuman power of thought, logic or language to create a radical outside. Again, it might seem obvious that Deleuze – in a manner explicitly different from his contemporaries – would not only forge a philosophy that mirrored theories of life that were not based on the coherence of individual units but would, more importantly, consider the very existence and possibility of thinking (and therefore philosophy) as a dynamic aspect of embodied and distributed mind. Here, his work would be different not only from the Foucault who sought to think the *pleasure* of bodies as modes of subject-formation *not* reducible to a milieu of life, and not only from a Derrida who would insist on the capacity of concepts to have a force above and beyond this world in a future to come, but also from Bernard Stiegler's politics that is grounded on epiphylogenesis (Stiegler 1998). For Stiegler, it is not only the case that human beings become what they are in relation to the world (including all the world's archives and technical objects), but that what a human being is, in general, is the outcome of a series of interactions with archives and technical objects. Despite the nomenclature, Stiegler's 'epiphylogenesis' is not only distinct from epigenesis, but tends to tame the radical potentiality of epigenesis. To speak of *epiphylogenesis* is to say that what is significant is some ongoing human kind or form which, for Stiegler, becomes an ongoing object of investment such that we can, *and should*, think of ourselves as more than life, as having a sublime relation to a future that we can only imagine. *Like* Deleuze, Stiegler draws heavily upon Simondon's theory of individuation where living beings are not unfolded from genomic profiles, but develop in relation to an equally dynamic milieu of other *folding* forces. What Stiegler adds to trans-individuation (and by way of his own individuating use of the archive) is a heightened and moral sense of human phylogeny: we, like all beings, are the result of the perception and retention of encounters, while the forces we encounter are also the consequences of dynamic relations that are sustained and transformed through time.

But to this sense of epigenesis Stiegler adds epiphylogenesis. What makes *human* becoming different is tertiary retention: we do not just retain, perceive and anticipate along with all the other living beings in the world; our written archive (of cave paintings, books, films) allows our interiority to already be a collective human exteriority. At its simplest we might say that rather than some type of potential expressing itself in relation to an outside that would then encourage the survival of some forms at the expense of others – a Darwinian statistical selection ranging over populations where it might still be appropriate to think of 'selfish' genes or some type of unit competing for survival against other possibilities – Simondon's conception of trans-individuation allows for the generation of an interior through complex and multiple encounters.

This is why Stiegler is able to think of trans-individuation as a way of thinking epigenesis that extends beyond its strict genetic definition. Rather than an interior finding its way in an exterior, or a living being becoming what it is in relation to its environment, there is one field of life or relations that generates stabilities; there are not beings who strive to survive, not individuals that are selected *at any level*, whether that be the gene or the organism. To say that something *is*, or for individuation to occur, is to speak of the folding of complexity, and a relation among forces. In the beginning is neither inside nor outside, neither body nor environment, *nor* a becoming in which there is a free and creative self-composition of something like 'life'. Neither Deleuze nor Stiegler adopt a theological model in which 'life' is the proliferation of as many rich and varied forms as possible, nor a mechanical model in which forms result from statistical variation and selection. An individual is not the preservation of a quality or possibility but the coming into being, through relation, of a complexity. A human individual is also the ongoing transformation of a difference in kind where, for Stiegler, phylogeny unfolds in relation to a milieu of texts and objects.

Where Stiegler retrieves and heightens the sense of human difference by describing epiphylogenesis, Deleuze – I want to argue – radicalises epigenesis by deploying the odd concept of preformationism. Where Stiegler's epiphylogenesis grants human life a distinct form and future by way of an archive that will allow for sublimation of our mere animality, Deleuze uses the concept of preformationism to situate an ideal, eternal and highly individuated power in life. For Deleuze – to use Stiegler's language – life already harbours tertiary objects: forms, powers, pure predicates that can operate across life, or at 'infinite speed' – not located in time and space but present virtually as if for all time. For Stiegler what epiphylogenesis amounts to, and what ought to concern us, is the fragility of human interiority. Being able to say 'I' does not result from reflecting upon one's subjectivity as a preceding condition – with the self becoming aware of itself as a transcendental condition – but is the effect of a history of reading and interiorising practices that create a self always in relation to other relations, and always reliant upon an archive. It follows *both* that the formation of this interiority is distinctly human (because of the formation of inscriptive technologies that allow for the retention of a past that enables the thought of complex and multiple futures), and that because of the distribution and exteriority of mind that our very interiority is exposed to risk. For Deleuze, by contrast, the precarious and contingent nature of human interiority is neither a reason for its urgent protection, nor anything like a justification for detaching 'thinking' or philosophy from life. Further, while Deleuze affirms what appear to be human philosophical powers – ideas, forms, pure predicates, the virtual and perhaps even concepts – these do not so much emerge from life as provide a better way of thinking about life in general as populated by forces of the outside.

This brings us back to what might happen to the relationship between life and philosophy if the notion of life is no longer focused on the autonomy of individuals. What concerns me here, therefore, is not so much the possibility of scientific realism per se, or whether phenomenology's inclusion of science *within* the lifeworld is legitimate, illegitimate or incompatible with realism. Rather, I want to put forward the notion of a general epigenesis that allows at once for the distinction of philosophical and scientific powers in life, but that neither opposes nor reduces philosophy to life. For Stiegler's epiphylogenesis the ongoing production of a distinct archive that is not reducible to day-to-day life – an archive that breaks with communication and easy transmission – is essential if the human is not to fall back into being the animal life from which (by way of epiphylogenesis) it has emerged. By contrast, much in Deleuze and Guattari's work suggests that the formation of the distinct organism, produced by relations with the forces of objects, is far from being felicitous and occurs only by way of misrecognising desire as that which disturbs organisms.

As I have already suggested, it would appear that the more theories of life move away from individuals or even individual units, such as genes, the more one would tend to think of minds as similarly always-already collective – as more like assemblages, collectives, populations or tribes than bounded self-maintaining organisms. It would also seem to follow that philosophy, accordingly, should focus less on what can be thought and less on freeing thought from life, and perhaps become more a question of the practices, institutions, relations and assemblages that compose technologies of the self.

Where does this leave us with regard to the concept of epigenesis in Deleuze's own work, and the significance – if any – of Deleuze's work for the present? One possibility is that if one accepts that Deleuze (and Deleuze and Guattari) not only endorse an epigenetic account of life but also allow philosophy to be vital rather than ideal or cerebral, then their work would be in line with a broader philosophical tendency to return mind to life, to demote thinking from its detached, Cartesian and (human) subjective elevation. Even if a theory of life cannot translate directly into an account of subjects and actions, it may nevertheless – as a concept – prompt a different conception of relations among forces. If that were so it would be necessary to mark out a difference between a literal translation of epigenesis into philosophy, and a philosophical or conceptual transformation. Whereas a literal acceptance of epigenesis would prompt philosophy to think of mind and reason as far more relational, but with an ongoing stability that affects and is affected by other forces, a consideration of life as governed by epigenesis might not necessarily reduce philosophy to a living practice, or transform its questions into problems of embodiment and relations.

As I have already suggested, most readings of Deleuze and Deleuze and Guattari tend to define their work as thoroughly in line with naturalised, materialist and embodied approaches both to the relation between knowledge and life, and to a new role and mode of philosophy (Braidotti 2002; DeLanda 2008; Grosz 2015). The first consequence of their uptake of a more symbiotic account of life would seem to be a transformation in how we understand knowledge practices: no longer subjects who are the condition for the possibility of knowing, but something closer to modes of existence and worlds that unfold from located points of view. This much seems to be accepted by Deleuze and Guattari, who refer to science's 'partial observers' (1994: 129): this does not mean that one relativises knowledge by seeing it as dependent on a subject, but that knowledge is generated by producing a perspective or point of observation that enables the formation of knowledge from a plane of reference. Scientific knowledge is impersonal, and yet subjective – where 'subjects' are produced through imagining what it would be to know, what it might be to occupy an observational possibility.

Here is where one might begin to see what is at stake in the relation between knowledge of life and what it is to think and, more importantly, to decide or act. For Foucault the normalising trajectory of biopolitics, where populations are managed on the basis of knowledge of life and increasing expertise, is only possible if one closes off questions of the ways in which knowledge practices are compositions or (to shift more toward Bruno Latour's slightly different take) modes of existence (Latour 2013). Despite their affinities, there's a marked difference between Foucault's principled critique of biopolitical normalisation and Deleuze and Guattari's account of stratification, and the relationship between thought and life. Not only do Deleuze and Guattari insist that desire (that which generates relations) rather than power (productive relations) should be the starting point (Deleuze and Guattari 1987: 531), and not only does

Deleuze criticise Foucault for staying inside relations, rather than thinking of life's radical 'superfold' (Deleuze 1988: 140). Deleuze and Guattari conclude with a far more impersonal and counter-subjective ethics than Foucault's theorisation of the ways in which bodies and practices might create formations irreducible to biopolitical and neoliberal reductions of existence to efficient managerialism.

To a certain extent, those who have theorised the shift toward epigenesis and its transformation of the ways in which one understands development in biology are explicit about the ways in which understandings of life emerge from a political context. In addition to work in the history of ideas that details the ways in which philosophers thought of epigenesis as a way of complicating the opposition between innate and acquired ideas, such that development was not merely a response to the world but retained the responsive history of the past (Malabou 2016; Mensch 2013; Muller-Sievers 1997), contemporary theorists of life, such as Scott Gilbert, have described the ways in which the twentieth-century theorisation of epigenesis emerged in the 1940s, where problems posed by Britain's role in the Second World War prompted a broader inquiry into forms of relation and creation: 'how does one coordinate different decisions so that they generate a coherent desired output? How do you go from parts to wholes? How does one mediate between competition and cooperation?' (Gilbert 2012). To say that theories of life are contextually bounded or determined is not the same as saying that a theory of life yields the way we should think about contexts. The concept of context – seeing any living being as made possible or inflected by its surrounds – already presupposes a theory of life that divides living being from milieu. To take Gilbert's example: to argue that Waddington's theory of the epigenotype was enabled by a social world that posed questions of cooperation reinforces a way of thinking about relations that a radical epigenesist would displace. Rather than relate scientific theories back to historical context, or to modes of existence or to politics – as though mind occurred *in life* – and rather than placing events such as philosophy or science back into the world, it might be possible – and I would argue desirable – to abandon inside/outside conceptions. To state that more clearly, the very notion of context – that how we think is part of a broader milieu of existence – is itself bound up with an understanding of life that is at once relatively recent in philosophy and that has an accompanying history of refusal.

When Foucault suggests that one might think about a body and its pleasures he is trying to detach sexual ethics from life, or trying to think of ethics other than through a sexual notion of the subject: subjects do not have 'a' sexuality that is somehow explicable in terms of life. To argue that there was a time when 'life' did not exist is obviously not to say that there were no living beings, but it is to say that there were bodies that were not referred back to a milieu. A 'history of bodies' would negotiate forces without any external, grounding, or prior plane. Subjects are practical formations, orientations toward one's body *not* as an object of knowledge, but as a possibility for experimentation. In this ethics of the subject Foucault is in part objecting to, and drawing upon, a Kantian mode of philosophy. Yet, one might claim that even when philosophy has placed life under erasure or out of play – in a tradition that runs from Kant to Foucault – a concept of individual life has nevertheless always been at play, both normatively and implicitly.

If life is seen to be something like a normalising ground (as it is in Foucault's conception of biopolitics) then it would be ethically and epistemologically urgent to suspend its capacity to close down questions of how one comes to know and posit life. If, however, the history of thinking about life epigenetically – where life is neither substrate, nor a principle of striving, survival or selection – disturbs the very notion of context, then the relation of bodies *to* context might require something more than a renewed relation between philosophy and life. The concept of epigenesis requires something like a different relation – a non-relation – between philosophy and life. What is placed in question is the way in which relations have been theorised as relations among bodies, or relations of bodies to life, lifeworlds and contexts. Let us consider a weak hypothesis: the relationship between philosophy and life depends very much on how we think of the life of the mind. If, to follow the school of embodied cognition, 'thinking' is not some detached computational activity that happens to take place in a body, but emerges from a complex network of responses, affects, habits and potentials, then a revolution in the scientific theory of life would necessarily have consequences for philosophy. More specifically, if one follows Henri Bergson, and argues that rather than taking its cue from the physical sciences (and quantitative mechanistic notions of matter) philosophy should follow the life sciences (where matter is not simply arranged but alters its nature depending on the ongoing matters with which it enters into relation), then a revolution in the science of life would be a revolution in the philosophy of life. Nothing, though, would have changed in the relation between philosophy and life. Philosophy would still rethink its own orientation according to a theory of life. If life presents itself in terms of stable bounded forms, then philosophy might follow a general organicism. If life is dynamic and self-organising then it would make sense to think of mind *in life* and as distributed and extended, rather than centred.

There is, however, another possibility, located in Deleuze's work on Leibniz, in Deleuze and Guattari's theorisation of the relation between philosophy and science, and in explicit and implicit moments in *A Thousand Plateaus*. When discussing epigenesis in relation to Leibniz, Deleuze raises the possible validity of preformationism, not so much to dismiss epigenesis or the power of development and unfolding, but to draw attention to folds in matter:

> It might be said that the theory of preformation and duplication, as observations made through the microscope confirm, has long been abandoned. The meaning of development or evolution has turned topsy-turvy, since it now designates *epigenesis* – the appearance of organs and organisms neither preformed nor closed one within the other, but formed from something else that does not resemble them: the organ does not arch back to a preexisting organ, but to a much more general and less differentiated design. Development does not go from smaller to greater things through growth or augmentation, but from the general to the special, through differentiations of an initially undifferentiated field either under the action of exterior surroundings or under the influence of internal forces that are directive, directional, but that remain neither constitutive nor preformative. However, insofar as preformism exceeds simple metric variations, it tends to be aligned with an epigenesis to the extent epigenesis is forced to hold to a kind of virtual or potential preformation. The essential is elsewhere; basically, two conceptions share the common trait of conceiving the organism as a fold, an originary folding or creasing (and biology has never rejected this determination of living matter, as shown nowadays with the fundamental pleating of globular protein). Preformism is the form in which this truth of the seventeenth century is perceived through the first microscopes. . . . From this point of view we cannot be sure if preformism does not have a future. (Deleuze 2001: 10–11)

Two phrases are worth emphasising here: 'a kind of virtual or potential preformation' and 'irreducible foldings'. Even though Deleuze is working through Leibniz, the rhetorical force of the book is to draw attention to a Baroque tradition – including the art and thought of the present – where interior and exterior are maintained in their difference, where there is a real distinction in an upper level. The 'future' of preformism hinted at here (and one that would be in line with epigenesis) is not one in which completed forms are given in advance, but one in which 'irreducible foldings' would preclude development from being the organisation or forming of undifferentiated matter. Matter bears an irreducible difference that *also* finds another level or articulation of difference as it unfolds. Further, the 'folds' in matter – or matter's tendencies, singularities or pure potentials – while folding into forms are *expressed* at another level in the ways in which those forms are perceived by other matters. More concretely, what is generated throughout the book on Leibniz is a difference between matter and manner. The difference is mutually constitutive, but the ways in which matter is expressed (manners) are different from the matters that offer themselves for expression. So, while Deleuze does – unlike many of his generation – refuse the shift in level of subjectivity, language or conceptuality that breaks radically with life, he nevertheless theorises a life that is not simply a plane of relations that produce relative stabilities. What occurs with stratification is both a forming of matters – that become substance through formation – *and* a distinct plane of expression. Here are Deleuze and Guattari speaking through Professor Challenger, where matter is said to harbour 'prevital' and *pre-physical* singularities, as though life, vitality and substance were not the random outcome of matter that has no difference in itself, but eventuated from matter that harbours distinct (but not yet differentiated) tendencies:

> He used the term matter for the plane of consistency or Body without Organs, in other words, the unformed, unorganized, nonstratified, or destratified body and all its flows: subatomic and submolecular particles, pure intensities, prevital and prephysical free singularities. He used the term content for formed matters, which would now have to be considered from two points of view: substance, insofar as these matters are 'chosen', and form, insofar as they are chosen in a certain order (substance and form of content). He used the term expression for functional structures, which would also have to be considered from two points of view: the organization of their own specific form, and substances insofar as they form compounds (form and content of expression). A stratum always has a dimension of the expressible or of expression serving as the basis for a relative invariance; for example, nucleic sequences are inseparable from a relatively invariant expression by means of which they determine the compounds, organs, and functions of the organism. To express is always to sing the glory of God. (Deleuze and Guattari 1987: 43)

Expression, or 'singing the glory of God', is the designation of a formed matter as *this*. The production of points of view, where matters form possibilities for an expression of the whole, generates multiple worlds (Deleuze 2001: 72). Ideality, the comprehension of the world, is unfolded from matter, but matter harbours folds that will create multiple *souls*. The soul is not some pre-existing interior, but it is also not simply an aspect of matter or point of view that surveys matter, as though the world as matter were relativised by its various ways of being perceived; the soul emerges as a really distinct expression, where points of view, souls or monads, are enabled by matter's own and different potentialities. Deleuze suggests that souls or monads 'having no windows' harbour their own styles or manners (4). Put more concretely, one might say that the powers of matter enable, prompt or call for quite different differences at the level of souls. This might be why art is so important not only in the book on Leibniz but throughout the Deleuzian corpus. Art is not – as it was for the French avant-garde – a negation of ordered and purposive life. For Deleuze, at least in his reading of Leibniz, the manners and styles in music and painting express what is distinct or *preformed* in matter. There is a preformationism in matter: not of simple units (such as genes or models) that unfold into fully fledged wholes, but of folds that allow for the variations of manners or styles. Matter's own complexities or folds are played out at a higher level. Works of art are not mental creations ex nihilo, pure inventions that require matter for their incarnation; they are responses to matter's infinite potentialities, and each expression is a singular point of view on the infinite. Truth, then, is not relative – mediated by the limits of finite perception – but there is a truth of the relative: all the folds in matter generate all the manners and styles of perception:

> Such is the basis of perspectivism, which does not mean a dependence in respect to a pregiven or defined subject; to the contrary, a subject will be what comes to the point of view, or rather what remains in the point of view. That is why the transformation of the object refers to a correlative transformation of the subject: the subject is not a subject but, as Whitehead says, a 'superject.' Just as the object becomes objectile, the subject becomes a superject. A needed relation exists between variation and point of view: not simply because of the variety of points of view (though, as we shall observe, such a variety does exist), but in the first place because every point of view is a point of view on variation. The point of view is not what varies with the subject, at least in the first instance; it is, to the contrary, the condition in which an eventual subject apprehends a variation (metamorphosis), or: something = x (anamorphosis). For Leibniz, for Nietzsche, for William and Henry James, and for Whitehead as well, perspectivism amounts to a relativism, but not the relativism we take for granted. It is not a variation of truth according to the subject, but the condition in which the truth of a variation appears to the subject. (21)

If artworks are neither purely subjective expressions nor straightforward copies, but are inflected by formed matters, with matter itself already being a creative variant of its singular and intrinsic difference, then how do those other events of mind – science and philosophy – relate to life. *What is Philosophy?* grants a difference to both science and philosophy that is compatible with Deleuze's own take on matter. What emerges is neither a hard philosophical naturalism where a scientific theory of life provides the foundation for philosophical method, nor a phenomenology where science's observations are conditional upon something like a general paradigm or way of seeing. The science of life – thinking of life in terms of the ways in which matters enter into relation and yield relatively stable modes of acting (or functions) – is the consequence of the creation of partial observers.

This is neither correlationism, where life is nothing more than the sense we make of it, nor hard realism, where science is the pure grasp of life in itself. It is, as I have already suggested a *philosopher's* way of thinking about life's capacity, in formed matters, to enable manners of thinking – either the functions of science or the concepts of philosophy. *What is Philosophy?* forms its own reflexive concepts of philosophy, science and art, in a manner of meta-philosophy. What it does not do is reduce philosophy's concepts to an adequate grasp of science's functions, nor see science's functions as direct unfoldings of the truth of matter. Where it does allow a theory of life to generate concepts of art, science and philosophy is in its understanding of the intrinsic difference of matter, and this in two senses. First, whatever matter is, it bears its own powers of formation; matter becomes what it is in foldings, or formed matters. In this respect there is some truth to preformationism insofar as what ultimately comes to be is not *fully* reducible to encounters and relations. Forms do not simply unfold, but they *are* the unfolding of matter's folds. To say that relations are external to terms (Deleuze and Guattari 1987: 391), is not only to say that no single being has – in advance – its own world that simply unfolds from some pre-given form; it is also to say that the whole that unfolds is really distinct. Second, if what Foucault criticised as biopolitics suffers from a literalism that would parse the supposed efficiency of life processes into a political managerialism, Deleuze does not respond by detaching arts of the subject from life, but grants life a complexity that is belied both by notions that there simply are organisms and individuals that enter into relation, and by conceptions of our understanding of life that would ideally bring philosophy and art into proximity with life's *own* generative power.

To talk about life's own generative power seems to be in line with the forms of new materialism that often invoke Deleuze's work. The coupling of preformationism and epigenesis in the book on Leibniz opens a slightly different way of thinking about life and genesis. If one talks about life as harbouring forms that are not simply models or blueprints but orientations, *and* if one then defines philosophy as at once distinct in its creation of concepts while also intuiting the composition of concepts in living forms, then the problem of genesis – of how thinking comes into being – is no longer about the emergence or transcendence of thought. Life is proto-philosophical in its harbouring of folds that generate differences, and philosophy is truly philosophical when it intuits life as almost or virtually ideal. Life and philosophy develop not so much in relation as in implication and explication. It is as though life awaits philosophy to intuit its conceptual potential, while philosophy must arrive at the point where it finds its concepts in life.

## References


Agamben, G. (1998), *Homo Sacer: Sovereign Power and Bare Life*, trans. D. Heller Roazen, Stanford: Stanford University Press.

Badiou, A. (2009), *Theory of the Subject*, trans. B. Bosteels, London: Continuum.

Bostrom, N. (2014), *SuperIntelligence: Paths, Dangers, Strategies*, Oxford: Oxford University Press.

Braidotti, R. (2002), *Metamorphoses: Towards a Materialist Theory of Becoming*, Cambridge: Polity.

DeLanda, M. (2006), *A New Philosophy of Society: Assemblage Theory and Social Complexity*, London: Continuum.

DeLanda, M. (2008), 'Deleuze, Materialism, Politics', *Deleuze and Politics*, ed. I. Buchanan and N. Thoburn, Edinburgh: Edinburgh University Press, pp. 160–77.

Deleuze, G. (1988), *Foucault*, trans. S. Hand, Minneapolis: University of Minnesota Press.

Deleuze, G. (2001), *The Fold: Leibniz and the Baroque*, trans. T. Conley, London: Continuum.

Deleuze, G. and F. Guattari (1983), *Anti-Oedipus: Capitalism and Schizophrenia*, trans. R. Hurley, M. Seem and H. R. Lane, Minneapolis: University of Minnesota Press.

Deleuze, G. and F. Guattari (1987), *A Thousand Plateaus: Capitalism and Schizophrenia*, trans. B. Massumi, Minneapolis: University of Minnesota Press.

Deleuze, G. and F. Guattari (1994), *What is Philosophy?*, trans. H. Tomlinson and G. Burchell, New York: Columbia University Press.

Derrida, J. (2003), *The Problem of Genesis in Husserl's Philosophy*, trans. M. Hobson, Chicago: University of Chicago Press.

Esposito, R. (2008), 'Totalitarianism or Biopolitics? Concerning a Philosophical Interpretation of the Twentieth Century', trans. T. Campbell, *Critical Inquiry* 34: 4, pp. 633–44.

Fink, E. (1933), 'Die phänomenologische Philosophie Edmund Husserls in der gegenwärtigen Kritik', *Kantstudien* 38, pp. 319–20.

Foucault, M. (1978), *The History of Sexuality, Volume 1: An Introduction*, trans. R. Hurley, New York: Pantheon Books.

Foucault, M. (2002), *The Order of Things*, London: Routledge.

Gilbert, S. F. (2012), 'Commentary: "The Epigenotype" by C. H. Waddington', *International Journal of Epidemiology* 41: 1, pp. 20–3.

Grosz, E. (2015), 'Deleuze and the Nonhuman Turn: An Interview with Elizabeth Grosz', *Deleuze and the Non/Human*, ed. J. Roffe and H. Stark, London: Palgrave, pp. 17–25.

Habermas, J. (1971), *Knowledge and Human Interests*, New York: Beacon Press.

Latour, B. (2004), 'Why Has Critique Run Out of Steam? From Matters of Fact to Matters of Concern', *Critical Inquiry* 30, pp. 225–48.

Latour, B. (2013), *An Inquiry into Modes of Existence: An Anthropology of the Moderns*, trans. C. Porter, Cambridge, MA: Harvard University Press.

Malabou, C. (2016), *Before Tomorrow: Epigenesis and Rationality*, trans. C. Shread, Cambridge: Polity.

Meillassoux, Q. (2008), *After Finitude: An Essay on the Necessity of Contingency*, trans. R. Brassier, London: Continuum.

Mensch, J. (2013), *Kant's Organicism: Epigenesis and the Development of Critical Philosophy*, Chicago: University of Chicago Press.

Muller-Sievers, H. (1997), *Self-Generation: Biology, Philosophy, and Literature Around 1800*, Stanford: Stanford University Press.

Protevi, J. (2009), *Political Affect: Connecting the Social and the Somatic*, Minneapolis: University of Minnesota Press.

Stiegler, B. (1998), *Technics and Time, 1: The Fault of Epimetheus*, trans. R. Beardsworth, Stanford: Stanford University Press.

Tragesser, R. (1984), *Husserl and Realism in Logic and Mathematics*, Cambridge: Cambridge University Press.

# Notes on Contributors

**Barry Allen** is Distinguished University Professor of Philosophy at McMaster University and a Fellow of the Royal Society of Canada. His books include *Truth in Philosophy*, *Art and Technology in Human Experience* and *Knowledge in Chinese Tradition*.

**Michael James Bennett** is Faculty Fellow in the Humanities at the University of King's College in Halifax, Nova Scotia. He is the author of *Deleuze and Ancient Greek Physics: The Image of Nature* (2017) and writes on the intersections between twentieth-century European philosophy and the history and philosophy of science.

**Claire Colebrook** is Edwin Erle Sparks Professor of English, Philosophy and Women's and Gender Studies at Penn State University. She has written books and articles on contemporary European philosophy, literary history, gender studies, queer theory, visual culture and feminist philosophy. Her most recent book is *Twilight of the Anthropocene Idols* (co-authored with Tom Cohen and J. Hillis Miller, 2016).

**Erin Hortle** is a Creative Writing PhD student at the University of Tasmania. An ongoing concern of her academic and freelance writing is the cultural inscription of the more-than-human world. Her work explores the ways in which creative writing helps us to understand how such inscriptions function, and how this, in turn, facilitates new ways of imagining, being and relating.

**Paul-Antoine Miquel** is Professor of Contemporary Philosophy at l'Université de Toulouse 2. He is the author of *Bergson ou l'imagination métaphysique* (2007) and *Sur le concept de nature* (2014).

**Johan Normark** is a researcher in archaeology, specialising in Maya Studies, in the Department of Historical Studies at the University of Gothenburg, Sweden. He has published numerous articles on the development of a non-anthropocentric archaeology, as well as his dissertation thesis *The Roads In-Between: Causeways and Polyagentive Networks at Ichmul and Yo'okop, Cochuah Region, Mexico* (2006). His research brings to bear the metaphysical insights of new materialism on archaeological issues of gender, ethnicity, perception, space, time, climate and landscape, among a host of other topics, practices and objects of various kinds. Johan blogs at *Archaeological Haecceities*.

**Tano S. Posteraro** is a PhD Candidate in the Department of Philosophy at Penn State University and Research Fellow at the Rock Ethics Institute. His dissertation reconstructs Henri Bergson's philosophy of evolution in order to update it for contemporary concerns. In addition to co-editing *Deleuze and Evolutionary Theory*, he has also published a number of articles on biological themes in Continental philosophy.

**Jon Roffe** teaches philosophy at Deakin University and the Melbourne School of Continental Philosophy. The co-editor of a number of volumes on twentieth-century and recent French philosophy, he is the author of *Badiou's Deleuze* (2012), *Abstract Market Theory* (2015) and *Gilles Deleuze's Empiricism and Subjectivity* (2016), and the co-author of *Lacan Deleuze Badiou* (2013) and *Practising with Deleuze* (2017). His next book is a collection of aphorisms, *Seduce or Die* (2018).

**Daniel W. Smith** is Professor of Philosophy at Purdue University. He is the translator of Gilles Deleuze's *Francis Bacon: The Logic of Sensation* and *Essays Critical and Clinical* (with Michael A. Greco), and the editor, with Henry Somers-Hall, of the *Cambridge Companion to Deleuze*. His book *Essays on Deleuze* was published by Edinburgh University Press in 2012.

**Hannah Stark** is Senior Lecturer in English at the University of Tasmania. She is the author of *Feminist Theory After Deleuze* (2017) and the co-editor of *Deleuze and the Non/Human* (2015) and *Deleuze and Guattari in the Anthropocene* (2016).

# Index

EU Authorised Representative:
Easy Access System Europe Mustamäe tee 50, 10621 Tallinn, Estonia
gpsr.requests@easproject.com

Printed and bound by CPI Group (UK) Ltd, Croydon, CR0 4YY
16/04/2026
02091725-0004